SpringerBriefs in Earth Sciences

For further volumes:
http://www.springer.com/series/8897

Alka Upadhyay · Raveendra Kumar Rai

Water Management and Public Participation

Case Studies from the Yamuna River Basin, India

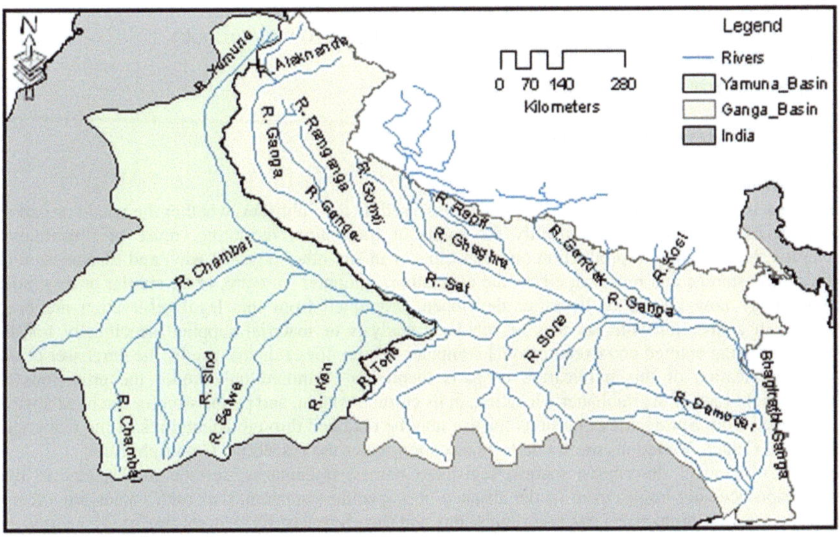

Springer

Alka Upadhyay
Water Resources Division
DHI (India) Water & Environment
Pvt Ltd
New Delhi
India

Raveendra Kumar Rai
Water Resources Division
DHI (India) Water & Environment
Pvt Ltd
New Delhi
India

ISSN 2191-5369 ISSN 2191-5377 (electronic)
ISBN 978-94-007-5708-0 ISBN 978-94-007-5709-7 (eBook)
DOI 10.1007/978-94-007-5709-7
Springer Dordrecht Heidelberg New York London

Library of Congress Control Number: 2012949713

Printed on acid-free paper

Springer is part of Springer Science+Business Media (www.springer.com)

Preface

Challenges faced by more and more countries in their struggle for economic and social development are increasingly related to water. Water shortages, quality deterioration, and flood impacts are among the problems which require greater attention and action. Globally the good quality of water for various uses has been scarce. Therefore, its management and allocation to meet out the various demand, such as domestic, agriculture, industrial, and environmental become important. However, the sustainable water resources management cannot be achieved without involvement of the stakeholders and public participation. This approach has been adequately addressed through the Integrated Water Resources Management (IWRM). Many counties have implemented this approach, but do not gain considerable success because of lack of public awareness and their involvement. In India, IWRM approach is being implemented by setting-up institution in the form of river basin organizations. For Yamuna river basin, effective management and planning of water resources accordance with geological formation, topography and climate has been investigated and linked to the public participation and awareness. The investigation was systematically planned and campaigning was made in different parts of the Yamuna river basin. Based on the analysis of information and available secondary data, necessary interpretations were made in light of the IWRM approach and documented as case studies in the present book.

This book has been presented into eleven chapters starting from the introduction to the water resources issues and brief idea of Indian water resources and principles of IWRM. Chapter 2 provides brief overview of the Yamuna River basin. The methodology adopted for the study is presented in Chap. 3. The Chaps. 4–10 presents the various case studies in different catchments of the Yamuna River basin. Finally overall, conclusion and lesson learned from the various case studies are summarized in Chap. 11.

The book will be useful to the water managers, decision makers, administrators, planners, Government, and Non-Government Organizations (NGOs) who involved in water resources management and poverty alleviation.

During the course of the campaigning, interactions was made with local people and are duly acknowledge for their timing and supports. Authors also wish to acknowledge local administration and local institutions for their co-operations and supplying available information. Authors wish to thank their colleagues for their necessary support.

Alka Upadhyay
Raveendra Kumar Rai

Contents

1 Introduction 1
1.1 Water Resources of India 2
1.1.1 Basin-Wise Per Capita Water Resources Availability of India 4
1.2 Water Management 6
1.2.1 Integrated Water Resources Management 9
1.2.2 Role of Public Partnership and Awareness 10
References 10

2 Brief Overview of the Yamuna River Basin and Issues 13
2.1 Issues and Challenges 15
2.1.1 Water Resources 15
2.1.2 Water Quality 18
2.1.3 Wastewater Generation 19
2.1.4 Environmental Flow 20
2.2 Summary of the Yamuna River Catchments 21
References 24

3 Methodology for Public Awareness Campaign 25
3.1 Conceptual Framework and Strategy 25
3.2 Selection of Villages 26
3.3 Primary Survey 26
3.3.1 Questionnaire Design 27
3.3.2 Group Interviews 27
3.4 Participatory Mapping 29
3.5 Secondary Data Collection 31
3.6 Data Analysis 31
3.6.1 Descriptive Analysis 31

3.6.2 Historical Analysis 31
3.6.3 Institutional Analysis 32
References 32

4 Case Study: Somb and Boli Catchments 33
4.1 Antiquity of Village 34
4.2 Geo-Physical Characteristics 34
4.2.1 Climate and Rainfall 34
4.2.2 Soil, Land Use and Land Cover 37
4.2.3 Biodiversity 37
4.2.4 Plantation 38
4.2.5 Common Crops 38
4.3 Inference from Primary Survey 39
4.3.1 Integrated Watershed Development Project 39
4.3.2 Environmental Status 40
4.3.3 Socio-Economic Status 42
4.3.4 Agriculture and Irrigation 43
4.4 Summary and Conclusions 45
References 50

5 Case Study: Middle Alluvial Plains 51
5.1 Geo-Physical Characteristics 53
5.1.1 Climate and Rainfall 53
5.1.2 Soil, Landuse and Land Cover 53
5.1.3 Biodiversity 54
5.2 Inferences from Primary Survey 54
5.2.1 Environmental Status 54
5.2.2 Socio-Economic Status 57
5.2.3 Amenities 59
5.2.4 Agriculture and Irrigation 59
5.2.5 Institutional Status 62
5.3 Summary and Conclusions 66
References 66

6 Case Study: Banganga Catchment 67
6.1 Geo-Physical Characteristics 68
6.1.1 Climate and Rainfall 68
6.1.2 Topography and Soil 68
6.1.3 Biodiversity 69
6.1.4 Common Crops and Cropping Pattern 69
6.2 Inferences from Primary Survey 69
6.2.1 Role of Local Institution 77
6.2.2 Environmental Status 77
6.2.3 Socio-Economic Status 79

6.2.4 Agriculture and Irrigation 80
6.2.5 Amenities 85
6.2.6 Institution 86
6.3 Summary and Conclusions. 86
References 90

7 Case Study: Banas Catchment 91
7.1 Geo-Physical Characteristics 92
7.1.1 Climate and Rainfall. 92
7.1.2 Topography and Soil. 92
7.1.3 Biodiversity 92
7.2 Inferences from Primary Survey. 92
7.2.1 Water Conservation 92
7.2.2 Local Institution 93
7.2.3 Environmental Status 94
7.2.4 Socio-Economic Status 95
7.2.5 Agriculture and Irrigation 96
7.2.6 Amenities 98
7.3 Institutions. 98
7.3.1 Capacity Building. 98
7.4 Summary and Conclusions. 102
References 102

8 Case Study: Chambal Catchment 103
8.1 Geo-Physical Characteristics 105
8.1.1 Climate and Rainfall. 105
8.1.2 Topography and Soil. 105
8.1.3 Biodiversity 105
8.1.4 Common Crops 106
8.2 Inferences from Primary Survey. 106
8.2.1 Water Conservation 106
8.2.2 Local Institution 108
8.2.3 Environmental Status 108
8.2.4 Socio-Economic Status 109
8.2.5 Agriculture and Irrigation 110
8.2.6 Amenities 111
8.3 Institution 111
8.3.1 Capacity Building. 115
8.4 Summary and Conclusions. 115
References 115

9 Case Study: Sind and Betwa Catchments 117
9.1 Geophysical Characteristics 118
9.1.1 Climate and Rainfall. 118

9.1.2 Topography and Soil. 118
9.1.3 Biodiversity 119
9.1.4 Common Crops 119
9.2 Inferences from Primary Survey 119
9.2.1 Water Conservation 121
9.2.2 Local Institution 124
9.2.3 Environmental Status 124
9.2.4 Socio-Economic Status 126
9.2.5 Agriculture and Irrigation 127
9.2.6 Amenities 131
9.3 Institution 132
9.3.1 Capacity Building. 133
9.4 Summary and Conclusions. 133
References 137

10 Case Study: Ken Catchment 139
10.1 Geophysical Characteristics 142
10.1.1 Climate and Rainfall. 142
10.1.2 Topography and Soil. 142
10.1.3 Biodiversity 142
10.1.4 Common Crops 143
10.2 Inferences from Primary Survey 143
10.2.1 Water Conservation 143
10.2.2 Local Institution 144
10.2.3 Environmental Status 144
10.2.4 Socio-Economic Status 145
10.2.5 Agriculture and Irrigation 146
10.2.6 Amenities 149
10.3 Institution 152
10.3.1 Capacity Building. 152
10.4 Summary and Conclusions. 152
References 152

11 Lesson Learned 153

Index 155

Chapter 1
Introduction

Abstract This chapter presents the status of water resources issue of India and the necessity of its management. The principles of integrated water resources management (IWRM) is also discussed for sustainability in the management including the role of public participation and awareness.

Water is essential for the socio-economic development and for maintaining the healthy ecosystems. It is a powerful geophysical force influencing the landscapes and climate. On a wide range of time scale, water is in everlasting movement, cycling through various reservoirs in the oceans, sky and soil. Indeed, water is abundant in the environment, but its distribution is quite uneven both in time and space. Therefore, to monitor the global water cycle with interaction among the key elements requires a comprehensive representation of the budgets and cycling of entire phases of water storages.

Reflecting on the water resources that will be available for future use, Hossain et al. (2011) have separated societal and environmentally important water resources into five categories: water food, energy, human health, and ecosystem function. These categories are inter-related and water is required for each of the remaining four categories. The water that would be available will depend on the demands or stressors placed on the water resources and the uses thereof. This principle applies on both quantity and quality of water resources. Some of the stressors may entail local human population requirements, irrigation, flood and drought, biofuel production, weather variability and long term change, land management practices, waste generation and treatment, animal and insect dynamics, vehicular and industrial emissions, natural landscapes change, natural events such as earth quakes and others. Among all, the major water stressor to the water resources is human population requirement, especially the food.

If considering the water requirement to feed the population, it is assumed that some 900 m^3 of water per person per year is required for self-sufficiency. However, Falkenmark (1997) has estimated this figure as 1,570 m^3 per person per year.

A. Upadhyay and R. K. Rai, *Water Management and Public Participation*,
SpringerBriefs in Earth Sciences, DOI: 10.1007/978-94-007-5709-7_1,

Considering the current global population, the estimated water required to feed the population is 7,117.5 BCM per year. However, this figure rises to 9,964.0 BCM per year for the projected population of 2050. It means, for the year 2050, the additional water requirement to feed the population will be 2,846.5 BCM per year. This is really a big future water challenge to be faced.

Beside population growth, climate change and landuse change will further aggravate the future water challenge, which results environmental damage and water sharing conflicts. Therefore, by now it is the prime importance to pay great attention to the water resources management. The only ways give the impression to resolve this issue is water resources management.

1.1 Water Resources of India

The water resources of Indian sub-continent largely depend upon the success of Indian Monsoon, especially the South–West (SW) Monsoon, generally experience during the month of June–September. It enters from the direction of Africa, and brings heavy rainfall to the west coast and covers large part of the northern India. It contributes around 80 % rainfall or precipitation in India. Other than the SW monsoon, during winter, average contribution of rainfall due to North–East (NE) Monsoon, which sweeps down from the plateaus of Asia and the Himalayas, and brings rain and cooler weather to south–east India between October and December. In spite of seasonal variability in the rainfall pattern, Indian sub-continent has large spatial variability in the mean annual rainfall/precipitation (MAP) ranging from 100 mm in Western Rajasthan to 2,500 mm in North–east part of the India with heaviest rainfall of approximately 11,000 mm near Cherapunji. High MAP values of approximately 2,000 mm are also typical to the western slopes of the Western Ghats.

This rainfall variability when coupled with geological and topographical conditions of the river basins, a large spatial variability in the flow regimes in the rivers experiences, which ranging from perennial to ephemeral. Based on the origin and flow regime, the rivers of India are classified as Himalayan, peninsular, coastal, and inland-drainage basin rivers. Himalayan rivers are snowfed and maintain a high to medium flows throughout the year. The heavy annual rainfall due to SW monsoon in the Himalayan catchments further increases their flow and caused flooding and erosion. These rivers are also fed with considerable base flow from groundwater in alluvial plains. Approximately 80 % of the total annual flows in the Himalayan Rivers are due to rainfall. The peninsular rivers on the other hand, are rainfed and do not have any snowmelt contribution. Approximately 90 % of the annual flows of these rivers are due to rainfall. The coastal rivers are found in the western part of India are short and episodic. Rivers of the Inland system (i.e. in the semi-arid to arid regions), centered in western Rajasthan, are few and frequently disappear in years of scant rainfall. These Inland Rivers are ephemeral in nature, and almost 100 % of its annual flows are due to the rainfall during SW

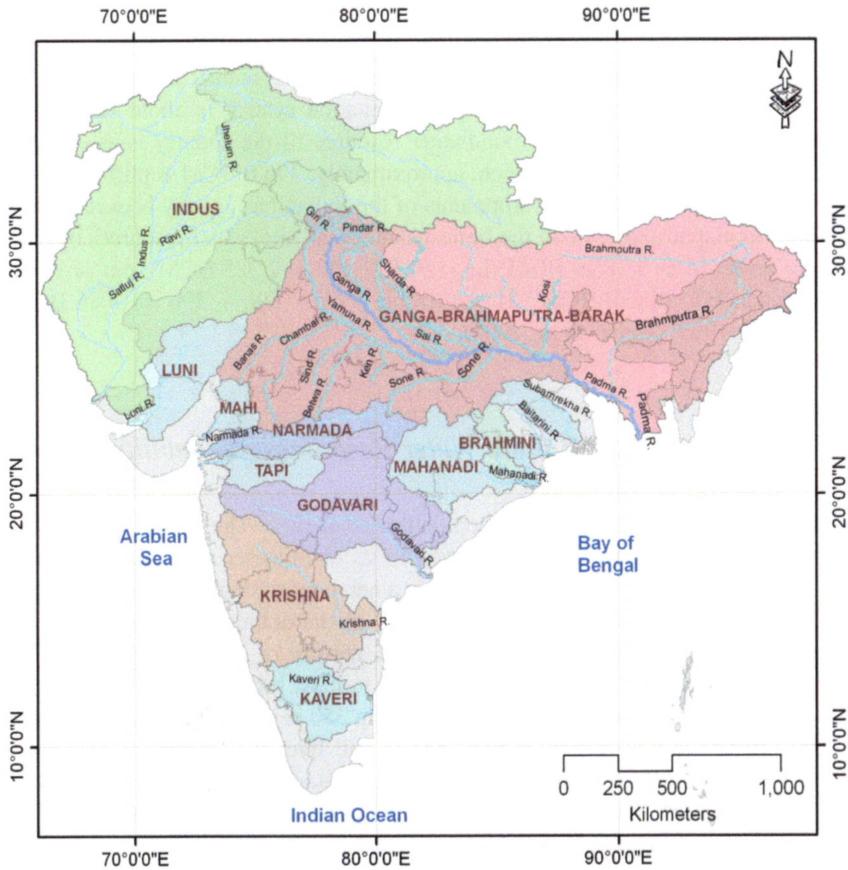

Fig. 1.1 Important river basins of India, including part of Pakistan

monsoon. The majority the rivers flow through broad, shallow valleys and drain into the Bay of Bengal. The river systems of India are shown in Fig. 1.1.

In the past, several organizations and individuals have estimated water availability for the Indian River basins. Recently, the National Commission for Integrated Water Resources Development estimated the basin-wise average annual flow in Indian River systems as 1,953.0 BCM (NCIWRD 1999). The utilizable water resource is the quantum of withdrawable water from its place of natural occurrence. Within the limitations of physiographic conditions and socio-political environment, legal and constitutional constraints and the technology of development available at present, utilizable quantity of water from surface flow has been assessed by various authorities. The utilizable annual surface water of the country is 690.0 BCM. In a majority of river basins, the percent utilization is quite high and is within the range of 50–95 % of utilizable surface resources, indicating a moderate to high availability of water for economic development, environmental conservation and social well-being.

Beside the surface water potential in the Indian basins, the annual potential natural groundwater recharge from rainfall in India is about 342.0 BCM, which is 8.56 % of the total annual rainfall of the country. The annual potential groundwater recharge augmentation from canal irrigation system is about 90.0 BCM. Thus, the total replenishable groundwater resource of the country is assessed as 432.0 BCM (CGWB). Out of which, approximately, 396.0 BCM is utilizable. The average stage groundwater developments of the Indian basins vary between 3.4 % in the Brahmaputra to 77 % in the Indus basin. Thus, total water resources of India can be assessed as 2,385.0 BCM (i.e. 1,953 + 432 = 2,385 BCM), out of which approximately 1086.0 BCM is utilizable [i.e. 690 + 396 = 1,086 BCM] (NCIWRD 1999; Kumar et al. 2005).

1.1.1 Basin-Wise Per Capita Water Resources Availability of India

In the basins, the distribution of population is uneven. The Ganga basin alone occupies about 40 % the total population in its one-quarter of the total drainage area of India. On the other hand, the five other largest single basins: Mahanadi, Brahmaputra, Krishna, Godawari cover 46 % of the drainage area but have only 30 % of the total population. The population density of India is 280 people per square kilometer, which is high as compared to most other developing countries. Six basins have population density of more than 350 persons per square kilometer.

Majority of the people in all river basins still lives in rural areas. More than 70 % of the Indian population is rural. This is substantially higher in some basins. For example more than 80 % of population is rural in Brahmaputra, Meghna, Mahanadi, Godavari and Brahmani-Baitarni basins. The livelihood of most rural population depends on agriculture. Thus the development and management of the available water resources is a crucial factor in the strategy of rural development and poverty alleviation in India.

To assess the adequacy of water resources availability, optimum limit considered is 1,700 m^3 per person per year. Based on this limit the water resources status may be categorized as: extreme scarcity: <500; scarcity: 500–1,000; stress: 1,000–1,700; adequate: 1,700–4,000; abundant: 4,000–10,000; and surplus: >10,000.

The basin-wise per capita water resources (PCWR) availability varies between 13,393 m^3 per annum for the Brahmaputra–Barak basin to about 300 m^3 per annum for the Sabarmati basin. Due to increasing population, water demand for various uses such as domestic, food, industrial and power will increase. The current national water demand is 813.0 BCM (GOI 1999). For the year 2050, NCIWRD (1999) has estimated that the total water demand in the country will be 973.0–1,180.0 BCM, whereas the utilizable surface and ground water resources are 1086.0 BCM. In spite of the spatial variability of utilizable water resources in the country, India will become a highly water stressed country, if better water management practices is not implemented at the watershed level. The PCWR

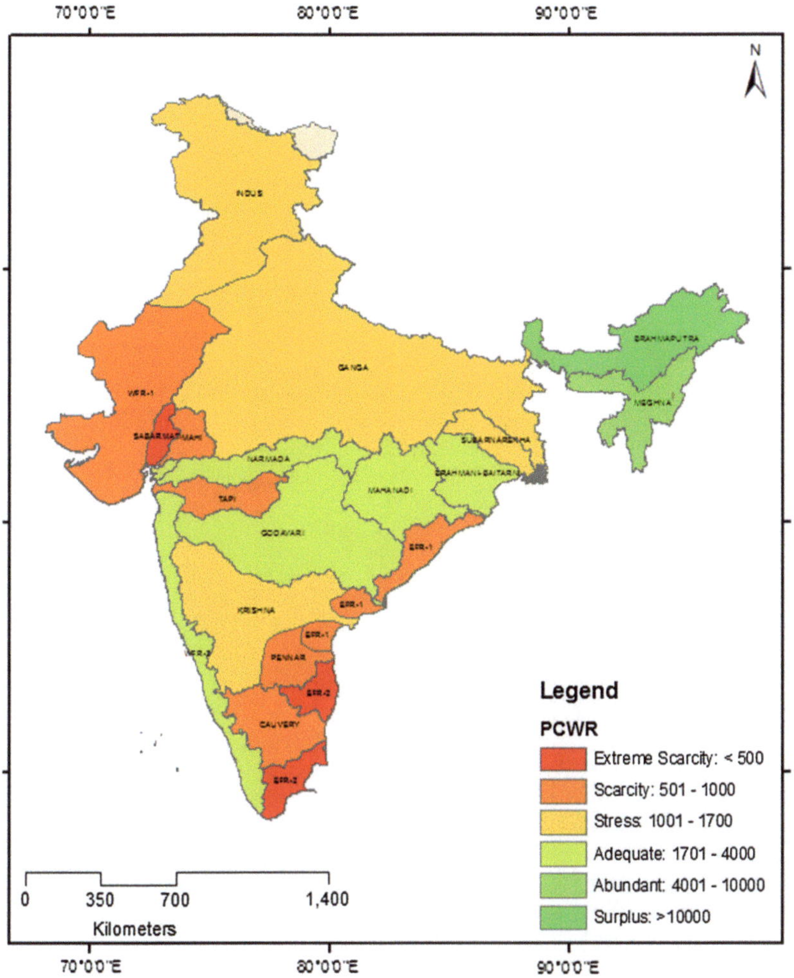

Fig. 1.2 Current per-capita surface water availability

analyses also reveal that only five basins have adequate water resources availability (Fig. 1.2). However, based on the analysis of per capita utilizable water resources potential (PUWR), currently three basins has the adequate utilization surface water potential (Fig. 1.3).

For the projected population of 2050, the per-capita surface water availability (PCWR-2050) is shown in Fig. 1.4. It is clearly apparent from Fig. 1.4 that India will face extreme water scarcity in future. Based on this analysis, it is observed that only Western Ghat Rivers, Brahmaputra and Meghna Rivers have adequate to surplus water resources availability.

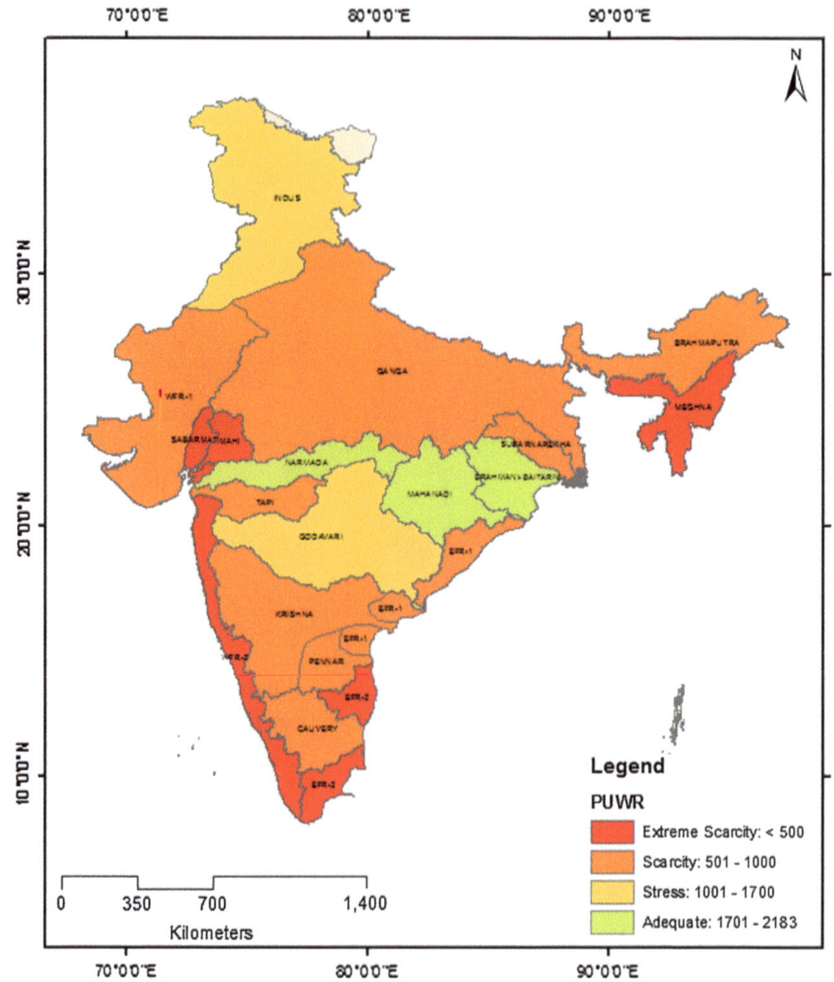

Fig. 1.3 Current per-capita utilization potential of water resources

Considering the above facts, it is necessary to manage the available water resources of India to suffice the water needs sustainably; which can be possible through integrated water resources management (IWRM).

1.2 Water Management

For sustainable water resources management, water accounting like inflows (precipitation, surface water inflow, ground water inflow), water use (evaporation, evapotranspiration, irrigation, drinking water), outflows (surface water outflow,

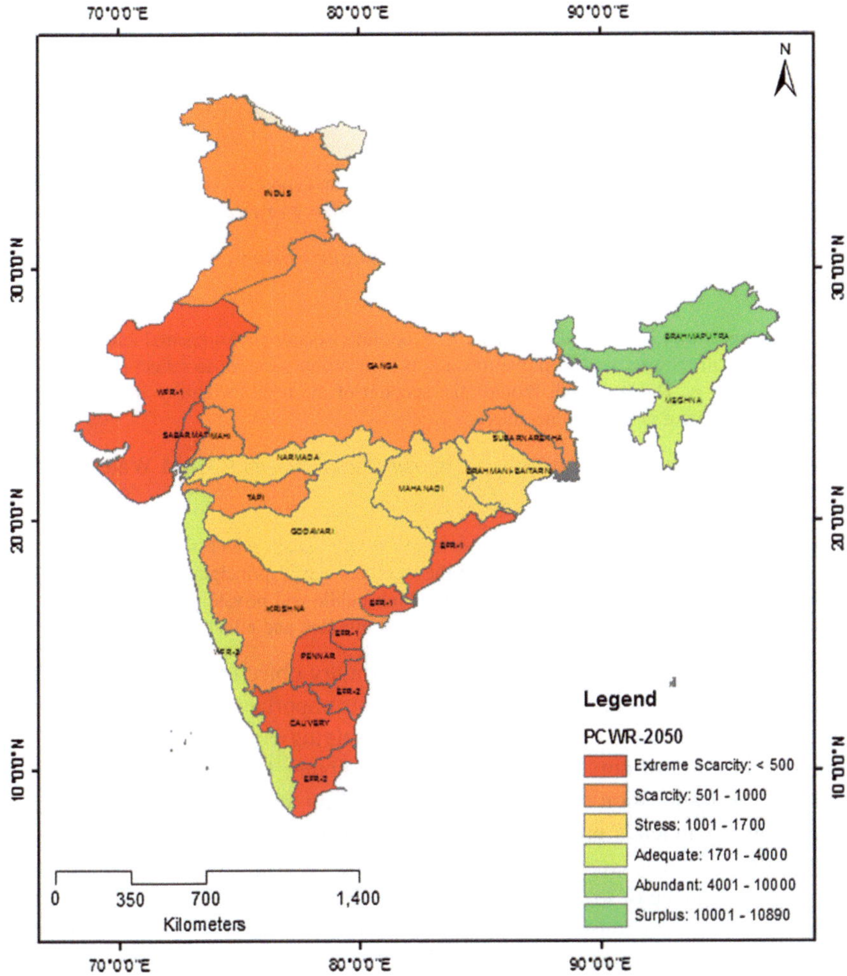

Fig. 1.4 Per capita surface water availability for the projected population

ground water out flow), and storage (surface storage, ground water storage, root zone storage) are the principal factors to be taken care of. The broad interventions for water management are: rainwater harvesting; groundwater recharge; maintenance of water balance; preventing water pollution; economic use of water; preventing various losses, etc.

For sustainable water resources management, a few important statements need to be carefully understood:

Dublin Statement on Water and Sustainable Development (ICWE 1992):

> Principle No. 4: Water has an economic value in all its competing uses and should be recognized as an economic good. Within this principle, it is vital to recognize first the basic right of all human beings to have access to clean water and sanitation at an

> affordable price. Past failure to recognize the economic value of water has led to wasteful and environmentally damaging uses of the resource. Managing water as an economic good is an important way of achieving efficient and equitable use, and of encouraging conservation and protection of water resources.

Agenda 21, *Chapter* 18 (UNCED 1992):

> Water should be regarded as a finite resource having an economic value with significant social and economic implications regarding the importance of meeting basic needs.

Ministerial Declaration of the 2nd *World Water Forum* (World Water Council 2000):

> To manage water in a way that reflects its economic, social, environmental and cultural values for all its uses, and to move towards pricing water services to reflect the cost of their provision. This approach should take account of the need for equity and the basic needs of the poor and the vulnerable.

Ministerial Declaration of the 3rd *World Water Forum* (World Water Council 2003):

> Funds should be raised by adopting cost recovery approaches which suit local climatic, environmental and social conditions and the "polluter-pays" principle, with due consideration to the poor. All sources of financing, both public and private, national and international, must be mobilized and used in the most efficient and effective way.

The focus on economic efficiency as the primary objective in the development and allocation of water resources is because of its importance as a social objective; efficiency values having viable meaning in resolving conflicts and assessing the opportunity costs of pursuing alternative uses (Young 1996). Although economically efficient allocation of irrigation water is rarely attained in practice, analysis of economic efficiency provides a useful point of reference for understanding the causes of inefficient allocation and mechanisms for improving the overall economic performance of irrigated production. As indicated, economic efficiency and equity are important considerations in the allocation of water. Greater efficiency is required in the face of increasing water scarcity, and equity is a concern because of the importance of water to the livelihoods and well-being of rural communities in particular. It is possible to derive a broad classification of policy measures that are relevant to managing resources within the boundaries of a nation. The measures include the redefinition of property rights and investment policies. One thing to be kept in mind is that many of the large public irrigation schemes that were promoted as part of the green revolution, particularly in India, were designed to target poor rural communities and as such were never oriented to maximize economic output, rather to guarantee of food security (Plusquellec 2002).

These principles of water management lead to integration of water, environment and social science, which further described as IWRM approach.

1.2.1 Integrated Water Resources Management

For sustainable water resources management, water science alone cannot provide concrete solution and understanding until it is related to the social and environmental activities. This integration associated with the cultural and spiritual perspectives is referred as the integrated water resources management (IWRM). The global water partnership (GWP) defines IWRM as follows (GWP 2009):

> IWRM is a process which promotes the coordinated development and management of water, land and related resources in order to maximize the resultant economic and social welfare in an equitable manner without compromising the sustainability of vital ecosystem.

Operationally, IWRM approaches involve applying knowledge from various disciplines as well as the insights from diverse stakeholders to devise and implement efficient, equitable and sustainable solutions to water and development problems. As such, IWRM is a comprehensive, participatory planning and implementation tool for managing and developing water resources in a way that balances social and economic needs, and that ensures the protection of ecosystems for future generations. An IWRM approach is an open, flexible process, bringing together decision-makers across the various sectors that impact water resources, and bringing all stakeholders to the table to set policy and make sound, balanced decisions in response to specific water challenges faced.

Considering the idea of IWRM and their implementation, necessary institutional set-up is required, and should comprise of various inter-disciplinary knowledge (Fig. 1.5). The scientists in various disciplines should have good communication and regular interaction with the stakeholders for holistic solution. The research and development activities in IWRM should be empowered by market based ideas, whereas water managers and stakeholders should be empowered by a foundation of knowledge.

Based on the various seminars, conferences and water weeks, three basic idea of IWRM were emerged, which are: (1) integrated water resources management in association with river basin management; (2) participation and stakeholder involvement; and (3) privatization/liberalization. The global consensus on the necessary shift in water governance has also been recognized as: (1) from administrative to resource based; (2) from centrally administrated to user based management institutions; and (3) from state to market-driven regulations.

In putting the IWRM principle into practice, many countries have adopted an approach where regulatory decisions such as water allocation and pollution licensing are implemented at the scale of the river basin or catchment. This has been accompanied by the emergence of institutional arrangements for water resources management that based on hydrological boundaries. While most of these institutions can be grouped as River Basin Organizations (RBOs) some are specifically mandated with managing groundwater aquifers and lakes basins. However, implementation of IWRM approach has not gain considerable success because of lack of sufficient institutional strength or power and public partnerships.

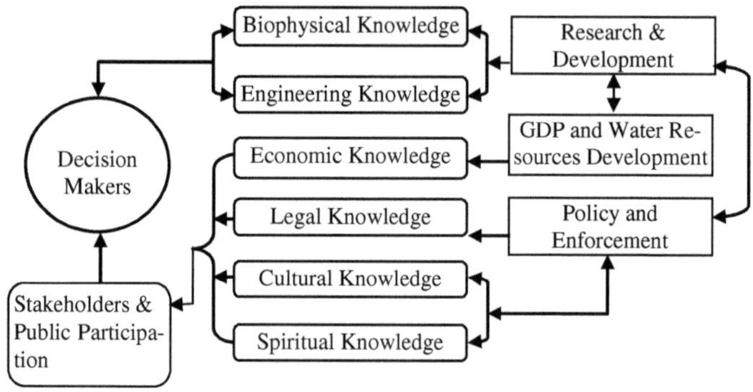

Fig. 1.5 Knowledge required for IWRM

1.2.2 Role of Public Partnership and Awareness

As stated above, institutional strength, public partnership and awareness are the key factor besides the biophysical and engineering understanding for the successful implementation of IWRM of river basin. However, public partnerships in the water management requires good insight of their natural resources (i.e. land and water), and climate so that they should aware about the pros-and-cons of the exploitation of natural resources with their economic benefits. The public awareness promotes more effective utilization and conservation of available water based on their landuse capability and climate.

Considering the significance of public partnerships and awareness, the detailed awareness campaign was carried out in the different parts of the Yamuna river basin, a sub-basin of river Ganga (India) to know the current status of water resources problems as well as the indigenous coping mechanism, and water related awareness. Based on the results of various case studies in different catchments of the Yamuna river basin, this book has been designed.

References

Falkenmark M (1997) Meeting water requirements of an expanding world population. Phil Trans Royal Soc Land 352:929–936

GOI (1999) Integrated water resources development. A plan for action. Report of the commission for integrated water resource development volume I. Ministry of water resources, New Delhi, India

GWP (2009) Global water partnership (http://www.gwptoolbox.org/)

Hossain F, Niyogi D, Adegoke J, Kallos G, Pielke Sr RA (2011) Making sense of the water resources that will be available for future use. Eos 90:144–145

ICWE (1992) The dublin statement on water and sustainable development. International conference on water and the environment, Dublin, Ireland on 26–31 Jan 1992

Kumar R, Singh RD, Sharma KD (2005) Water resources of India. Curr Sci 89(5):794–811

NCIWRD (1999) Integrated water resources development—a plan for action, report of the national commission for integrated water resources development, ministry of water resources. New Delhi

Plusquellec H (2002) How design, management and policy affect the performance of irrigation projects. Emerging modernisation procedures and design standards. FAO, Bangkok

UNCED (1992) The Earth Summit: Agenda 21, Chapter 18: protection of the quality and supply of freshwater resources—application of integrated approaches to the development, management and use of water resources. United Nations conference on environment and development, Rio de Janeiro, 3–14 Jun 1992

World Water Council (2000) Ministerial declaration of the Hague on water security in the 21st century. Second world water forum on 22nd Mar 2000

World Water Council (2003) Ministerial declaration of the 3rd world water forum in Koyota in 2003

Young RA (1996) Measuring economic benefits for water investment and policies. World Bank Technical Paper No. 338, The World Bank, Washington

Chapter 2
Brief Overview of the Yamuna River Basin and Issues

Abstract This chapter presents the current status of the Yamuna river basin as well as brief description of its catchments. This chapter also includes the various issue related to water in terms of water quantity, quality, wastewater and environmental flow.

The Yamuna River is one of the important and sacred rivers of India. It is the largest tributary of the River Ganga. It originates from Yamunotri glacier in the Mussoorie range of the lower Himalayas, and after traversing 1,376 km joins the river Ganga at Allahabad. The drainage area of the Yamuna basin is 366,220 sq km, which comprises part of seven states, viz. Uttarakhand, Himachal Pradesh, Uttar Pradesh, Haryana, Delhi, Rajasthan and Madhya Pradesh (Table 2.1). The Yamuna River has four main tributaries in the Himalayan region: Rishi Ganga, Hanuman Ganga, Tons, and Giri. In the plains, the main tributaries are the Hindon, Chambal, Sind, Betwa and Ken (Fig. 2.1).

The river water is generally used for irrigation, drinking and industries as well as for mass bathing, laundry, cattle bathing, and secretion of the cremation ash. The construction of diversion structures at regular intervals (Hathinikund, Wazirabad, Okhla, Gokul, etc.) for irrigation, domestic and industrial water supply, has largely modified the flow regime of the river. The inflow of wastewater either treated or partially treated in the river further aggravates the water quality problem of the river. Though the green revolution was important for food security, but lack of regulation in the groundwater abstraction has led to ground water table depletion causes damage in causal linkage between surface and ground water, resulting change in surface water dynamics during the lean season of the river. This is the main reason of dry river segments observed between Hathinikund and Palla (Delhi).

A. Upadhyay and R. K. Rai, *Water Management and Public Participation*, SpringerBriefs in Earth Sciences, DOI: 10.1007/978-94-007-5709-7_2,

Table 2.1 Details of catchment area and state of Yamuna river basin

State	Area in Yamuna basin (km^2)	Area in major sub-basin (km^2)					
		Hindon	Chambal	Sind	Betwa	Ken	Others
Uttarakhand	3,771 (1.1 %)	–	–	–	–	–	3,771 (3.74 %)
Himachal Pradesh	5,799 (1.7 %)	–	–	–	–	–	5,799 (5.76 %)
Uttar Pradesh	70,437 (20.4 %)	7,083 (100 %)	452 (0.32 %)	748 (2.89 %)	14,438 (30.12 %)	3,336 (13.66 %)	44,380 (44.06 %)
Haryana	21,265 (6.1 %)	–	–	–	–	–	21,265 (21.11 %)
Rajasthan	102,883 (29.7 %)	–	79,495 (56.87 %)	–	–	–	23,388 (23.22 %)
Madhya Pradesh	140,208 (40.6 %)	–	59,838 (42.81 %)	25,131 (97.11 %)	33,502 (69.88 %)	21,090 (86.34 %)	647 (0.64 %)

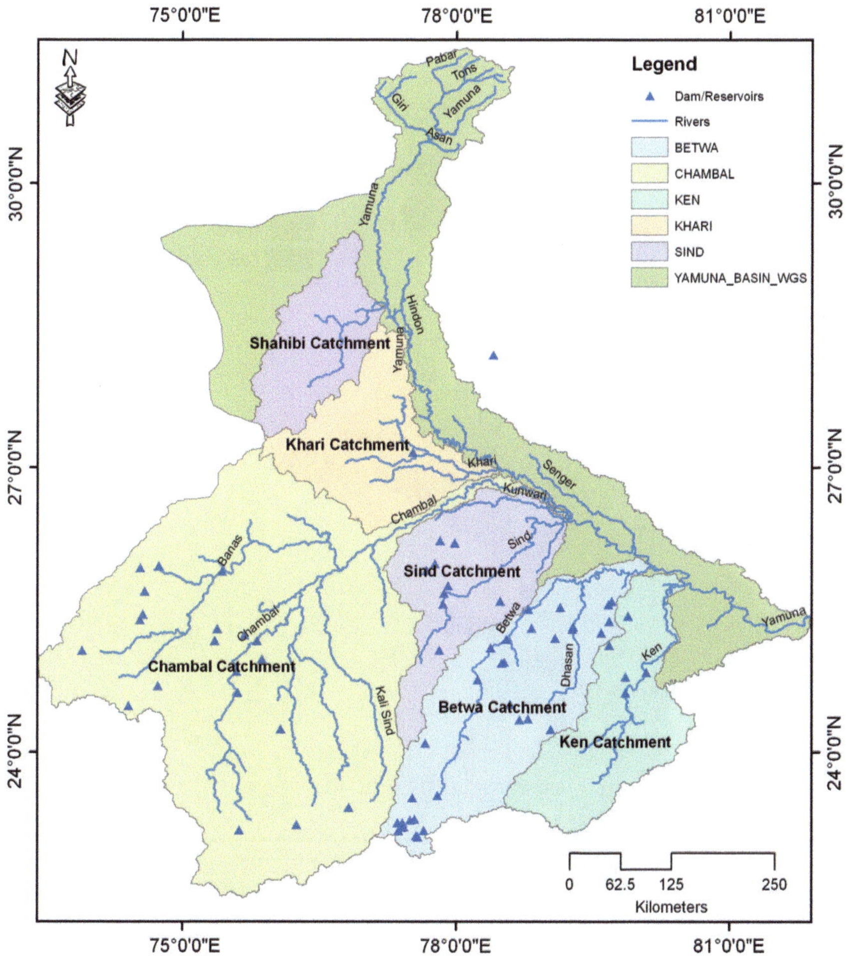

Fig. 2.1 Catchments of the Yamuna river basin and major tributaries

2.1 Issues and Challenges

The major issues and the challenges of the Yamuna river basin may be categorized as water resources, water quality and wastewater generation. These issues are briefly elaborated in the following sections.

2.1.1 Water Resources

Since independence, the population in the Yamuna basin has been increased by 2.5 times, and to ensure the food security, various irrigation schemes have been subsequently developed. Currently, the irrigated area in the basin has become 2

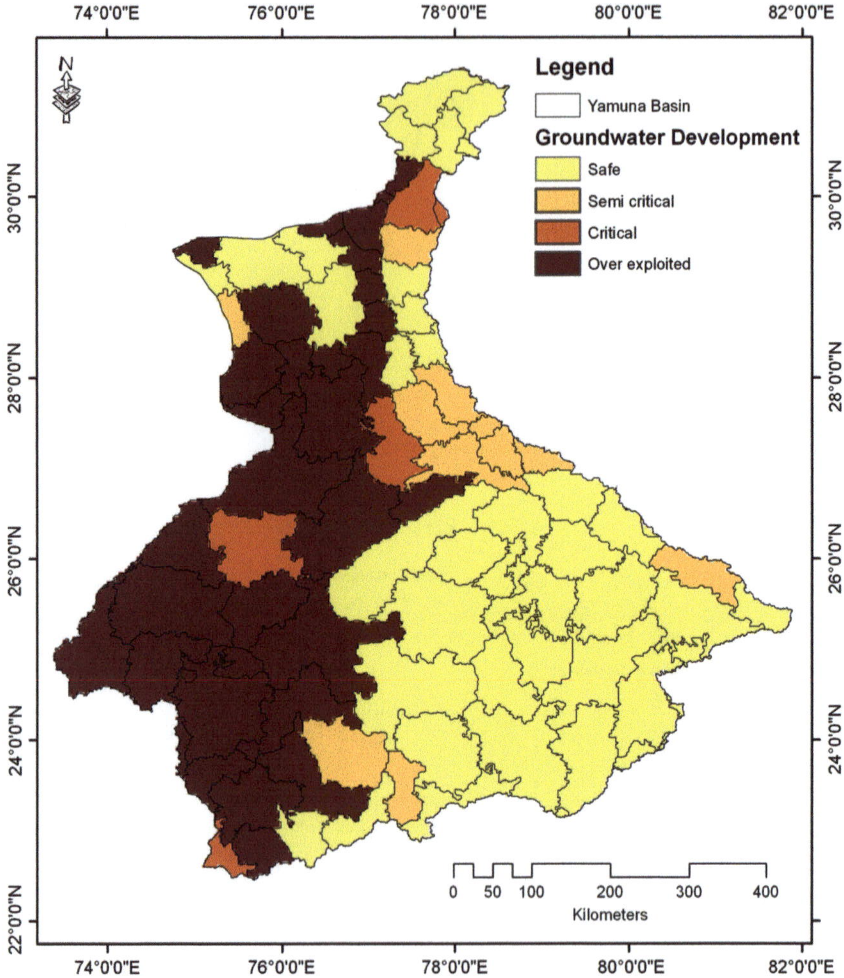

Fig. 2.2 Stages of ground water development in the basin (Rai et al. 2011)

times as compared to the year 1950. Also, during 60 years since 1950, the contribution of groundwater resources for irrigation has been triplicated. A net share of surface and groundwater for irrigation has now been become 37 and 57 % instead of 60 and 20 % in 1950s. The rest of the irrigation water is met by other sources like ponds, tanks, *anicuts*, etc.

The increasing groundwater development in the basin is a major concern in sustainable water resources planning and management. The district-wise stage of ground water development in the basin is depicted in Fig. 2.2, which clearly indicates that eastern and south-east part of the basin has already been overexploited in terms of CGWB classification (CGWB 2010). The right bank portion of the main Yamuna River is also under semi-critical to overexploited category of

Table 2.2 Mean monthly flows of Yamuna river at various locations (Rai et al. 2011)

Months	Mean monthly flows (m³/s)				
	Paonta	Kalanaur	[a]DRB	[a]Agra	Pratappur
June	84.9	35.9	37.8	27.0	356.8
July	302.4	287.8	248.3	157.2	2,426.8
August	595.5	604.8	630.9	505.9	6,608.4
September	399.3	389.6	453.3	504.6	6,230.7
October	104.5	89.6	93.3	157.2	1,166.7
November	53.2	52.1	57.8	81.0	569.0
December	40.7	11.9	35.8	43.6	466.7
January	40.7	11.9	34.3	29.8	366.2
February	56.2	10.3	33.5	28.9	392.6
March	71.7	43.7	28.0	20.9	345.3
April	30.8	17.7	42.4	27.0	356.8
May	40.3	17.2	24.3	17.4	242.6

[a] Though the flow is available but water quality is very poor

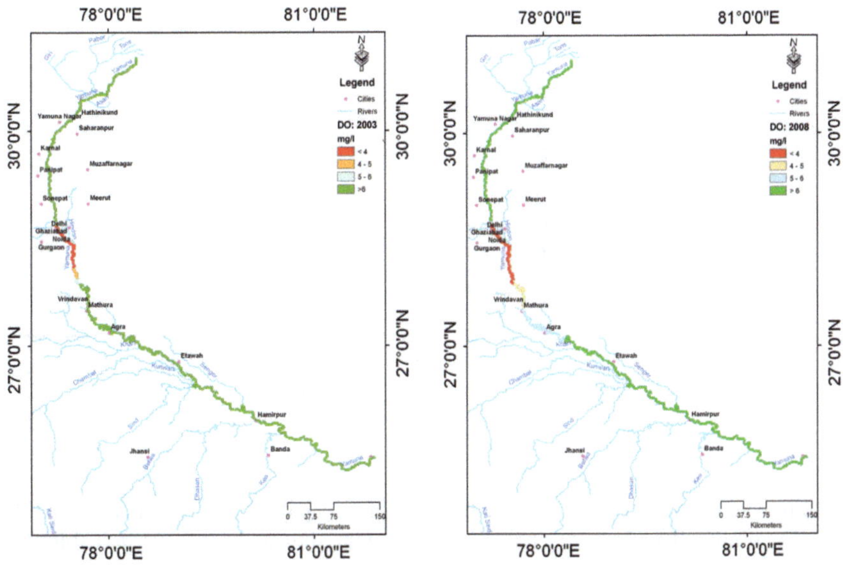

Fig. 2.3 Comparison of water quality profile for the year 2003 and 2008 (Rai et al. 2011)

stage of ground water development starting from Yamunanagar to Etawah, this led to termination in the causal linkage between the groundwater and the river runoff.

The current water use and the analysis of 10 years (2000–2009) of hydrological data (both surface water and groundwater) reveal that entire basin is under water deficit (approximately 34.54 BCM of deficit). However, the Ken River (surplus of 3.27 BCM) and the Upper Himalayan (surplus of 4.6 BCM) catchments have

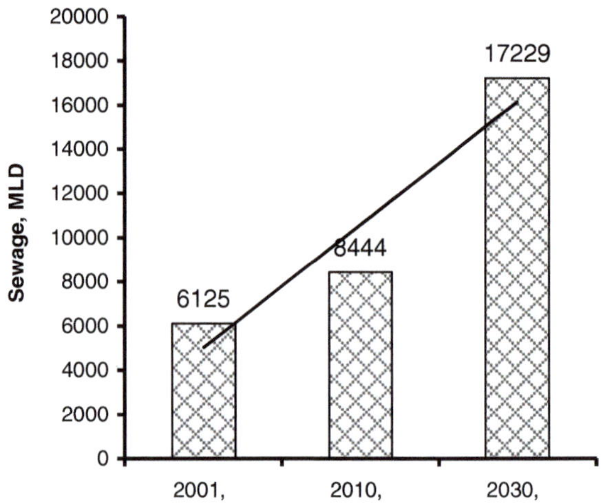

Fig. 2.4 Projection of wastewater generation

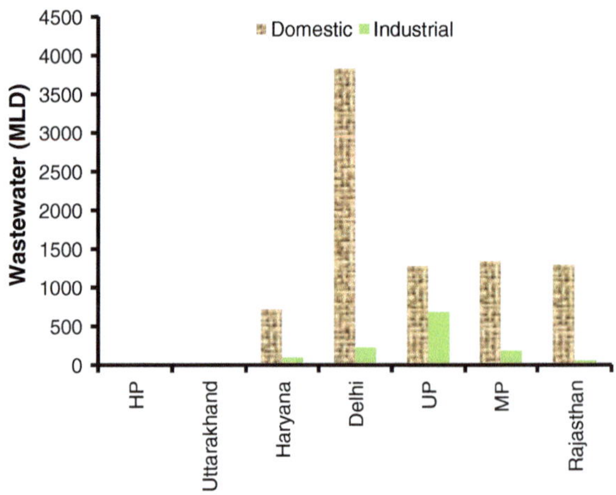

Fig. 2.5 State-wise domestic and industrial wastewater

surplus water resources because of its poor land capabilities. The mean monthly flows available at various discharge measurement sites are specified in Table 2.2.

2.1.2 Water Quality

Analysis of the secondary data collected from Central Pollution Control Board (CPCB) and primary sampling at various locations in the Yamuna River reveals that water quality is gradually deteriorating and will tend to deteriorate if not managed properly. This can be seen through the comparison of DO profile of the Yamuna River for the year 2003 and 2008 (Fig. 2.3), which indicates that polluted stretch in the Yamuna River, is gradually increasing.

Table 2.3 Wastewater generation and treatment capacity of YAP towns

Cities/towns	Bank	Volume of WW (MLD)	Treatment capacity (MLD)
Yamunanagar	R	45	35
Saharanpur	L	45	38
Muzaffarnagar	L	40	32.5
Karnal	R	60	48
Panipat	R	60	45
Sonepat	R	45	30
Delhi	R/L	3,800	2,330
Gurgaon	R	45	30
Faridabad	R	140	115
Ghaziabad	L	150	126
Noida	L	90	70
Vrindavan	R	5	4.5
Mathura	R	35	28
Agra	R	190	90
Etawah	R	13	10
Allahabad	L	223	89
Total		4,986	3,121

2.1.3 Wastewater Generation

Wastewater inflow in the river Yamuna is major source of pollution, which is governed by population, water supply, sewerage network and collection, efficiency of the relevant infrastructure, etc. Currently, 8,444 MLD of wastewater is generated in the basin, out of which about 4,458 MLD is discharged directly into the Yamuna river and about 1,200 MLD is discharged into its tributaries remaining 2,786 MLD is either disposed of on land or used for irrigation. Due to population growth, this wastewater generation will be further aggravated (Fig. 2.4). Delhi alone generates about 3,743 MLD of wastewater, which is 44 % of the entire sewage generated in the basin and 84 % of the sewage being discharged into the Yamuna along its entire course.

State-wise wastewater generation is shown in Fig. 2.5, which demonstrates that the Uttar Pradesh is highest contributor of industrial waste in the Yamuna River through Hindon River and Hindon-cut.

On the sewage treatment front, Delhi has highest sewage treatment capacity of 2,330 MLD which is about 68 % of the total sewage treatment capacity in the basin (Table 2.3). However, there is still a large gap between generation and treatment in Delhi itself. Though, several STPs have been installed along the river course with a designed capacity of the order of 2,332.25 MLD for Delhi, 327 MLD for Haryana, and 403.25 MLD for Uttar Pradesh. But, the river quality of the river is not yet improved as the treatment capacity is neither adequate nor effectively utilized. Main issues related to sewage management are:

Table 2.4 Environmental flow requirement of various reaches of Yamuna river

Month	Release from Hathinikund (Hathinikund to Palla) (m^3/s)	Release from Wazirabad (Wazirabad to Okhla) (m^3/s)	Release from Okhla (Okhla to Agra) (m^3/s)	[a]Net release from Hathinikund for maintaining the environmental flow up to Agra (m^3/s)	From Agra (Agra to Etawah) (m^3/s)	From Etawah (Etawah to Allahabad) (m^3/s)
January	36.8	14.5	42.4	82.2	63.8	91.6
February	40.4	14.7	45.5	89.9	67.8	98.8
March	47.2	12.3	36.7	78.4	54.9	86.3
April	37.0	17.6	48.4	88.0	69.7	89.2
May	39.2	11.0	33.8	73.8	51.1	60.7
June	46.4	15.99	45.5	89.5	66.6	89.2
July	121.8	83.3	117.3	188.5	196.8	606.7
August	227.2	189.0	221.1	318.0	435.4	1,707.3
September	155.6	162.4	209.5	323.6	409.0	1,507.5
October	63.0	36.4	115.3	167.6	182.4	301.4
November	48.5	22.4	69.4	113.8	107.4	137.7
December	36.8	15.0	47.0	87.0	72.9	116.7

[a] The net release accounts the ecological flow of previous reach which will carry forward to the next reach

- STPs capacity is inadequate as compared to the generated sewage.
- STPs are in general practically not meeting their compliance.
- Under running of most of the STPs due to lack of sewer connections.
- Improper drainage system.
- Excess BOD concentration coming to the plant due to inadequate water supply, etc.

Considering the current status of generation and treatment capacity of Delhi alone, even if treated effluent quality is achieved at 10 mg/l BOD for the entire existing treatment capacity, still BOD load would be 179 t/d, which may result in BOD concentration of 46 mg/l in the final effluent and not complying with the prescribed standards. Even if the entire sewage of Delhi is treated to a level of 5 mg/l of BOD, still the BOD load in the final effluent would be 19.4 t/d, which may continue to impair the water quality of the river.

2.1.4 Environmental Flow

Environmental flow is an important issue for the river Yamuna, and an attempt was made to estimate it for different stretches in different months. In the study, various hydrology based methodology was investigated. A hydraulic method (Rai et al. 2011), which accounts for seepage and evaporation losses, ecological requirement and water requirement for pollution assimilation was used to estimate the

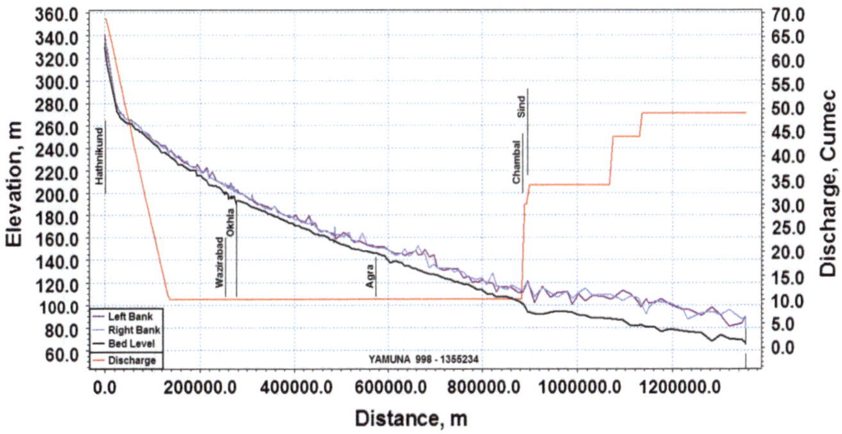

Fig. 2.6 Flow profile of the Yamuna with Hathinikund release of 70 m^3/s

stretch-wise environmental flow requirement. The estimated minimum environmental flow release from the Hathinikund barrage without considering the dilution water for pollution assimilation was 73.8 m^3/s (Table 2.4), which is approximately equal to the 50 % of mean minimum flow available at Hathinikund.

The base line environmental flow was also simulated using the MIKE 11 model without considering the water resources development in the river Yamuna, and it was appeared that approximately 70 m^3/s of flow is required to be released (i.e. minimum release) from the Hathinikund barrage to maintain the ecological flow of the river up to Etawah (i.e. before the confluence of river Chambal (Fig. 2.6). However, this quantity of water for maintaining the ecological flow is really a great challenge because of other priority uses of water such as drinking and irrigation.

2.2 Summary of the Yamuna River Catchments

Based on the analysis of climate, topography, landuse, geology, population status, etc., problems and their possible solution is summarized in Table 2.5.

Considering this large spatial variation of the water related problems/issues in the basin (Table 2.5), the public awareness and partnership campaign was planned for different catchments. The methodology adopted and the case studies are presented in subsequent chapters.

Table 2.5 Summary of the physiographic and climatic conditions of the sub-basin

Variables	Upper Himalayan	Hindon	Hathinikund to Delhi
Mean rainfall (mm)	1175	887.0	650.0
Climate	Humid	Semi-arid to sub-humid	Semi-arid
Mean slope (m/km)	28.528	<0.70	<0.80
Hydrogeology	Hilly and Tarai	Recent alluvium	Recent to older alluvium
Geology	Soft rock and Tarai	Alluvium	Alluvium
Groundwater development	White	White to grey	Grey to dark
Cropping intensity (%)	118–167	132–167	148–212
Population density (per sq km), census-2001	12–300	600–2000	100–800
Population density along the river	Population density along the river course is varying between 800 and 9,412; which causing major environmental damage to the river.		
Economy	Horticulture + Agriculture	Agriculture + Industries	Agriculture + Industries
Pollution sources	Domestic Industries (Scattered source)	Industries Domestic	Domestic Industries
Major water problem	–	Quality	Quantity and quality
Solution	Construction of environmental reservoir for riparian ecosystem maintenance Soil conservation	Wastewater management Crop planning Improvement in irrigation efficiency	Crop planning Improvement in irrigation efficiency Groundwater recharge Wastewater management River training work for water augmentation through weirs across the river

Variables	Chambal	Sind	Betwa	Ken
Mean Rainfall (mm)	783.7	848.3	1064.9	1125.0
Climate	Semi-arid to sub-humid	Semi-arid to sub-humid	Semi-arid to moist sub-humid	Sub humid to humid
Mean Slope (m/km)	1.704	1.066	1.063	1.5527

(continued)

Table 2.5 (continued)

Variables	Chambal	Sind	Betwa	Ken
Hydrogeology	Older alluvium Consolidated sedimentary Crystalline (Igneous and metamorphic) Deccan trap	Older alluvium Consolidated sedimentary Crystalline (Igneous and Metamorphic) Deccan trap	Consolidated sedimentary Crystalline (Igneous and metamorphic) Deccan trap	Older alluvium Consolidated sedimentary Crystalline (Igneous and metamorphic)
Geology	Hard rock	Hard rock	Hard rock	Hard rock
Groundwater development	Mostly dark	White to grey	White to grey	White to grey
Cropping intensity (%)	105–212	105–132	105–167	105–148
Population density (per sq km), census-2001	100–300	100–600	100–600	100–600
Economy	Agriculture + Livestock	Agriculture + Livestock	Agriculture + Livestock	Agriculture + Livestock
Pollution sources	Domestic (Scattered source)	Domestic (Scattered source)	Domestic (Scattered source)	Domestic (Scattered source)
Major water Problem	Quantity	Quantity	Quantity	Quantity
Solution	Crop planning Improvement in irrigation efficiency Soil conservation Water harvesting and groundwater recharge through shaft	Crop planning Improvement in irrigation efficiency Soil conservation Water harvesting and groundwater recharge through shaft	Crop planning Improvement in irrigation efficiency Soil conservation Water harvesting and groundwater recharge through shaft	Crop planning Improvement in irrigation efficiency Soil conservation Water harvesting and groundwater recharge through shaft

References

CGWB (2010) Groundwater status of India. Central Ground Water Board, Faridabad

Rai RK, Upadhyay Alka, Ojha CSP, Singh VP (2011) The Yamuna river basin: water resources and environment. Water science and technology series, vol 66, Springer, The Netherlands

Chapter 3
Methodology for Public Awareness Campaign

Abstract This chapter presents the methodology and approach adopted for conducting the public awareness and partnership campaign in the basin. Focus group discussion at different levels and discussion with women has been conducted beside the questionnaire filling. The secondary data from local institutions were also collected to confirm the information received during the focus group discussion. The officers of the local government institutions like Patwari, Gram Pradhan, Village Guard, etc., were also consulted. The information was analyzed and the inferences were drawn and scaled up for the catchment.

As described in Chap. 2 that the Yamuna River basin comprised an area of approximately 361,000 km^2 and has very complex hydrologic system due to its great heterogeneity in hydro-geological settings and climatic behaviour. Therefore, to plan such a study, a conceptual framework and strategy need to be clearly demarcated. The criteria of selecting the village on sample basis was given importance so that whole basin can be adequately addressed.

3.1 Conceptual Framework and Strategy

Conceptual framework was designed according to the following two fundamental issues:

1. *Pathways of Development* comprising reason and implications associated with natural resource management and poverty alleviation.
2. Rural development and resource conservation *Policies and Programs* appropriate to cope with the trade-offs in the critical triangle of sustainable watershed development in the region and to what extent can they facilitate and enhance sustainable rural developments.

A. Upadhyay and R. K. Rai, *Water Management and Public Participation*, SpringerBriefs in Earth Sciences, DOI: 10.1007/978-94-007-5709-7_3,

3.2 Selection of Villages

The primary criteria for villages identification was based on: (1) the water stress conditions either due to quantity or quality, and (2) any running or abandoned water management schemes or institutional set-up for water management. The selection of villages (or micro-watersheds) was done through a *purposive sampling*. The villages were selected that fulfils most of criteria, such as: (a) topographic dissimilarities, (b) towns identified under Yamuna Action Plan, (c) industrial town area, (d) groundwater development and exploitation, (e) water quality and quantity related sensitivity among masses (f) water conservation: traditional/modern practices, (g) water user associations, etc.

Further, to know the overall status in the basin, a cluster technique was adopted to regionalize the approach at sub-basin level. The number of village considered for the study was based on the area of catchments and their heterogeneity. So that general recommendation can be scaled up to catchment scale.

3.3 Primary Survey

The aim of primary data collection is to identify different patterns of change and to study their causes and effects on the catchment's socio-economic response and environment due to any anthropogenic developments. The extensive village survey has an overall objective of the collecting primary information about the current status of available resources, water management activities or initiatives, water quality, waste disposal practice, cropping pattern, and other land use changes. Also, it was tried to cater "the past and present status of the resources, and coping mechanism under adverse conditions". It was also attempted to find out the opinion of villagers regarding the climatic changes and their impacts on cropping patterns. Primary survey also focused on the following other driving forces:

1. to describe **key changes** in the study region including shift factors, outcome variables, implementation of policies and programs, changes in local perceptions, organizations and actions in natural resource management; water quality and availability;
2. to analyze local-level **determinants of technological and institutional change** and their outcomes in terms of economic, ecological and social environments;
3. to understand the nature of **collective and individual responses** of villages to the pressures of change including their corresponding implications for the economic and ecological conditions;
4. to derive **policy recommendations** that facilitates potential benefits for people and nature through watershed development in specific pathways and restricts their negative consequences.

The actual survey of the villages was conducted through the structured questionnaire and participatory research techniques.

3.3.1 Questionnaire Design

The questionnaire was designed to cater all the information relating to water, environment and society. Questionnaire includes the following aspects (Tables 3.1, 3.2):

- General: population (growth, composition, average household size); change in health; education; infrastructural indicators; welfare changes; drinking water, waste disposal, sanitation and hygiene, etc.
- Agriculture and forestry: land use changes and their causes for agricultural and forest land; change in agricultural practices; government interventions and effects; land use intensity and changes; introduction of new products and technologies; irrigation.
- Natural resources and their management: perception of changes in soil quality, water availability and quality, forest cover and forest quality changes and biodiversity.

3.3.2 Group Interviews

The principal instruments of the survey were group interviews using a group questionnaire to produce oral histories of changes in the village and about the perception of the villagers of these changes. Interviews were conducted with key informants to collect quantitative and qualitative data on the key variables. Interviews were conducted separately with the villagers and *Gram Pradhan*, *Patwari*, local members and the water user association (WUAs) followed by group interview. *The central role of women in water resources management and the need for women to be more involved in decision-making at all levels clearly informs the Dublin Principles* (ICWE 1992), *and was also cited, inter alia, in Principle* 20 *of the* 1992 *Rio Declaration, in the* 1995 *Beijing Platform for Action, in the* 2002 *Johannesburg Plan of Implementation, and in resolution* 58/217 *of the General Assembly, which proclaimed* 2005–2015 *as the International Decade for Action, 'Water for Life (*www.gdptoolbox.org)*'*. Therefore, to know the exact view of the overall status of the life, a separate interview was also performed with the women. A crucial point was the involvement of older people into the study as the focus is on historical changes over the last 2 decades.

Primary information for drinking water status, irrigation, cropping pattern, crop production, land holding, fertilizer application, pollution, diseases, living standard, biodiversity, waste disposal, migration, village institutions, education, status of women and coping mechanism were discussed through participatory research appraisal by focus group discussion with village elders, village institutions head man and women.

Table 3.1 Questionnaire for focus group discussion for water management practices

S. No.	Item	Remarks
1	Water conservation and agriculture practices	
	(i) What conservation practices for water are common in the area?	
	(ii) How many ponds, WHS are there in your village	
	(iii) Case of ponds encroachment in your area and purpose:	
	(iv) Do you get sufficient water from canal: Rabi: Kharif: Zayed	
	(v) Name of distributaries/minor command in which your village fall in:	
	(vi) Cases of un authorized canal cutting/breaching in your area:	
	(vii) What is cropping pattern in village?	
	(viii) Do you get sufficient water for: Rabi: Kharif: Zayed	
	(ix) Irrigation facility (a) Canal: (b) Tube well Power source Pump used (centrifugal/submersible) (c) Water harvesting structure/ponds (d) Any other	
	(x) What average production do you get from agriculture: Rabi: Kharif: Zayed:	
	(xi) What are common vegetables grown in your village?	
	(xii) What are common fruits grown in your village?	
2	Fertilizer and pesticides	
	(i) Use of fertilizer: Consumption (per ha), type/brand	
	(ii) Organic farming Use of pesticides: Consumption (per ha), type/brand	
	(iii) Bio-pesticides	
	(iv) Practice of integrated pest management	
	(v) Practice of integrated nutrient management	
3	Waste disposal method	
	(i) Filling in low land area	
	(ii) Surface disposal	
	(iii) Any other	
4	Toilet facility in village	
	(i) Open defecation	
	(ii) Community toilet/in house facility	
5	Water quality and quantity	
	(a) Drinking water facility/source (HP/dug well/tube well/pond/any other)	
	(b) Water level: Pre monsoon, post monsoon	
	(c) Drinking water quality (adequate/not adequate)	
	(d) Irrigation water quality	

(continued)

Table 3.1 (continued)

S. No.	Item	Remarks
6	Water borne diseases	
	(i) Pathogenic	
	(ii) Non pathogenic	
7	Farming/forestry scheme in village (government/NGO sponsored)	
8	Water conservation/watershed management scheme on village (government/ NGO sponsored)	
9	Flora and fauna	
	(i) Type of trees/shrub in the area:	
	Type of trees/shrub in the area before 20–25 years:	
	Type of grasses in the area:	
	Type of grasses in the area before 20–25 years:	
	(ii) Fauna in the area:	
	Animal	
	Reptile	
	Birds	
10	Best indigenous practices in the area (water conservation. organic farming, social forestry)	
11	Have you seen any industrial waste disposal (solid/water) in your area?	
	Name of industry/type of waste	
12	YAP-I has been completed. Current YAP-II and previous YAP-I plan has contributed in your area: yes/no… if yes	
	Do you know Yamuna action plan, who is doing/what is going on in this:	
	Do you find any change in WQ in your area since last 10 years?	
	Does the water availability for drinking water have changed in last 10 years?	
	Does the water availability for irrigation have changed in last 10 years?	

3.4 Participatory Mapping

Spatial data is critical to grasp the interactions between resource use, agricultural production and human welfare on a specific site. Therefore, participatory mapping (Emmel 2008) was undertaken in the selected villages to visualize resource use changes and to facilitate the understanding of these changes over time according to villager's perceptions. Together with complementary secondary data and photographs, as these enable to identify the pathways of development and specify the hypotheses on the induced innovation of technological change and its impacts on land use and resource condition. Geo-referenced data and information were collected prior to the village survey in order to get site-specific information about administrative units, watershed boundaries and topography.

Table 3.2 Questionnaire for focus group discussion for drinking water

S. No.	Parameters	Remarks
1	Availability of water	
	a Round the year	
	b 8 months (except summer months)	
	c 4 months (only during monsoon months)	
2	Adequacy of water	
	a For drinking, household need and livestock	
	b For drinking and household needs	
	c Only for drinking and cooking	
	d Overall rating (sufficient/insufficient)	
3	Source of drinking water	
	a Open well/bore/hand pump/underground tank/regional scheme/others	
	b Drinking water arrangement for livestock (cattle trough/pond/at farm/ others)	
4	Average distance of public source from house	
	a <100 m	
	b 100–200	
	c 200–500 m	
	d >500 m	
5	Average daily time spent by a family for accessing water during summer months/opportunity cost of achieving water security	
	a Up to 2 h/day	
	b 2–4 h/day	
	c >4 h/day	
6	Dependency on temporary measures for water security during last 5 years (in number of years)	
	a Dependence on government machinery	
	b Dependence on private water tankers/private bore well/hand pump etc	
7	Quality of available drinking water	
	a (Potable/hard/brackish/saline/fluoride/nitrate/turbid/other problem)	
	b Any water related diseases found in village in past/present?	
8	Socio-economic condition	
	a Day to day water conflicts due to water scarcity in summer months (frequent/uncommon/none)	
	b Average monthly expenditure for getting domestic water	
	c Day to day water availability for marginalized section of society	
9	Current management/maintenance of the water supply system/source	
	a Who does it? (government/panchayat/others)	
	b Management (good/average/poor)	
	c Maintenance (good/average/poor)	
	d Water charges collected	
	e Do villagers pay water charges (if yes, fully/partial)	
	f Water committee formed in village?	
10	Impact of watershed initiatives	
	a For irrigation (good/average/poor)	
	b For drinking water security (good/average/poor)	
	c Women's position in the watershed programme (notable/insignificant)	

3.5 Secondary Data Collection

On the village level, secondary data were used for indicators, which might be difficult to assess and to estimate from the villagers. Secondary data provide the possibility to crosscheck the data collected through primary survey. Secondary data for the 3–4 crop calendar years were collected to see the pattern change. Combined investigation of primary and secondary data on block/village level will enable decision makers to obtain a deeper understanding of the process of change, such as reconstruction of the economic, resource history, possible effects on economic, socio-cultural and ecological variables, etc.

3.6 Data Analysis

The collected primary or secondary data were subjected to analysis for interpreting the impact of any institutional set-up for water management, public participation, development of water management schemes, etc. For this purpose, both drinking water supply augmentation and crop productivity were analysed for pre- and post interventions. Three types of analyses were performed: (1) descriptive, (2) historical and (3) institutional.

3.6.1 Descriptive Analysis

Computations and comparisons were conducted for data from secondary and primary sources. Subjective viewpoints and perceptions of the villagers on indicators as resource quality, welfare changes over time, changes in land use, biodiversity using qualitative data analysis.

3.6.2 Historical Analysis

A number of different instruments were used to provide the historical dimension of the research. Secondary data of population, agriculture and welfare were evaluated at village level. In the survey, oral history techniques were applied to obtain a picture of economic and environmental history of villages.

An attempt was also made to investigate/discover the sequences of change and their dominant causal linkages between the different factors. The quality of the recall data depends crucially on the questioning techniques of the researcher and on the selection of participants in the group interview, e.g., across social classes and gender.

3.6.3 Institutional Analysis

Key institutional indicators are the local organizations related to land use, water use, and forest/pasture use. These organizations help in regulating and managing the existing government programs and external donors in the field of natural resource management. The aim of the institutional analysis is to characterize the effect of local institutions and organizations on resource use patterns.

Based on the analysis made through the aforementioned approach, status of water management practices and water quality in the basin was identified at watershed scale, and planning and management program can be structured on the priority basis. Using the above methodology of public partnership and awareness campaign, selected villages were investigated in various catchments of the Yamuna river basin, and are presented as case studies in subsequent chapters.

References

ICWE (1992) The Dublin statement on water and sustainable development. International Conference on Water and the Environment, Dublin, Ireland

Emmel N (2008) Participatory mapping: an innovative sociological method. Real Life Methods-Sociology, University of Manchester, Manchester

Chapter 4
Case Study: Somb and Boli Catchments

Abstract This chapter presents the case study of Somb and Boli catchments of the Yamuna River basin. The area has overexploited the ground water resources, but water conservation techniques have little improved the water problems, but still need integrated watershed management practices. Public has played significant role in the watershed development programmes initiated by the government and NGOs, but vocational training for off-season income, and water related awareness will further improve the operational part of the development. Capacity building is another aspect, which need extra emphasis.

The Somb and Boli catchments are located in the Shivalik foothills of the Yamuna river basin (Rai et al. 2011). The campaigning for water management and public partnership issues were conducted in the two villages of these catchments. These villages were Dhanaura (*Hadbast No*. 252) and the Nagli (*Hadbast No*. 32) located at the upper part of the Shivalik foothills of Bilaspur block of Jagadhari tehsil of Yamuna Nagar district. Jagadhari tehsil is an important industrial town located at the left bank of the Yamuna River, which is famous for metal, utensil and plywood industries. Jagadhari comes under the over-exploited zone category in terms of ground water (CGWB 2010). Ground water pollution is another threat in the Tehsil demanding adequate attention. A major canal system originated from the Hathi-nikund barrage [i.e. Western Yamuna Canal (WYC)] is passing through the tehsil. Because of disposal of treated industrial wastewater from five major industries in the area, water pollution in the WYC and the Yamuna River was initiated. However, a drain up to Radaur was constructed, which will discontinue this practice of wastewater disposal into the canal, and enhance the wastewater degradation time before mixing into river Yamuna. The industries bring cash flow to the town and development in all aspects such as road, communication, school, hospitals and employment. However, beneficiaries of the industries are centered in urban place, and villagers always get out of the periphery of the industrial benefit.

A. Upadhyay and R. K. Rai, *Water Management and Public Participation*,
SpringerBriefs in Earth Sciences, DOI: 10.1007/978-94-007-5709-7_4,

This has been a general tendency in the developing countries, and of course the Jagadhari is not an exception.

In spite of the above industrial development, villages of Jagadhari have set an exemplary model for water resources augmentation and management through participatory approach. Rural inhabitant are not only surviving because of these interventions but have paused population migration towards the urban, which otherwise creates slum in urban areas.

The Dhanaura is situated at the foothills of Shivalik range of Himalayas and river Somb passes through it. It comes under Ranjitpur *halka* of Bilaspur block; whereas, Nagli is situated at the upper reaches and river Boli is passes through it. It comes under Manakpur *Halka* of Bilaspur block. The Somb and Boli River merges at Dadupur and finally joins the river Yamuna at Meharmajra village. The location map of the habitation in the villages is shown in Fig. 4.1.

4.1 Antiquity of Village

In the early 90s, for reclaiming wasteland under the direction of English officers some work was done to plant out the sandy wastes along the present or former beds of the streams with trees and grass, with the objective of reclaiming waste land and protecting cultivated areas from erosion. *Sarkdna* grass (*Saccharum sara* or *Ciriale*) and *kdhi* (*Saccharum spontaneum*) are invaluable for this purpose, and many thousands of acres were reclaimed in this way. The grasses are valuable in themselves for rope-making and other numerous village purposes, but their principal use is that they bind the soil and prevent the extension of damage by sand, while they require little to encourage their growth beyond combined efforts on the part of the villagers and simple measures for protection from the cattle.

The most important of the hill terrains include Somb and Boli River in Jagadhri. The bed is covered with large boulders for a few miles below the hills, but soon becomes a wide tract of sand. The upper portion of the course contains water throughout the year a foot deep in summer, but reaching six feet or more during monsoon causing flood with dangerous current.

A current general overview of these villages can be made through the photographs taken during the visits (Fig. 4.2).

4.2 Geo-Physical Characteristics

4.2.1 Climate and Rainfall

The climate of the area is sub-tropical, characterized by a hot and dry summer, southwest monsoon season and a bracing cold season. The southwest Monsoon

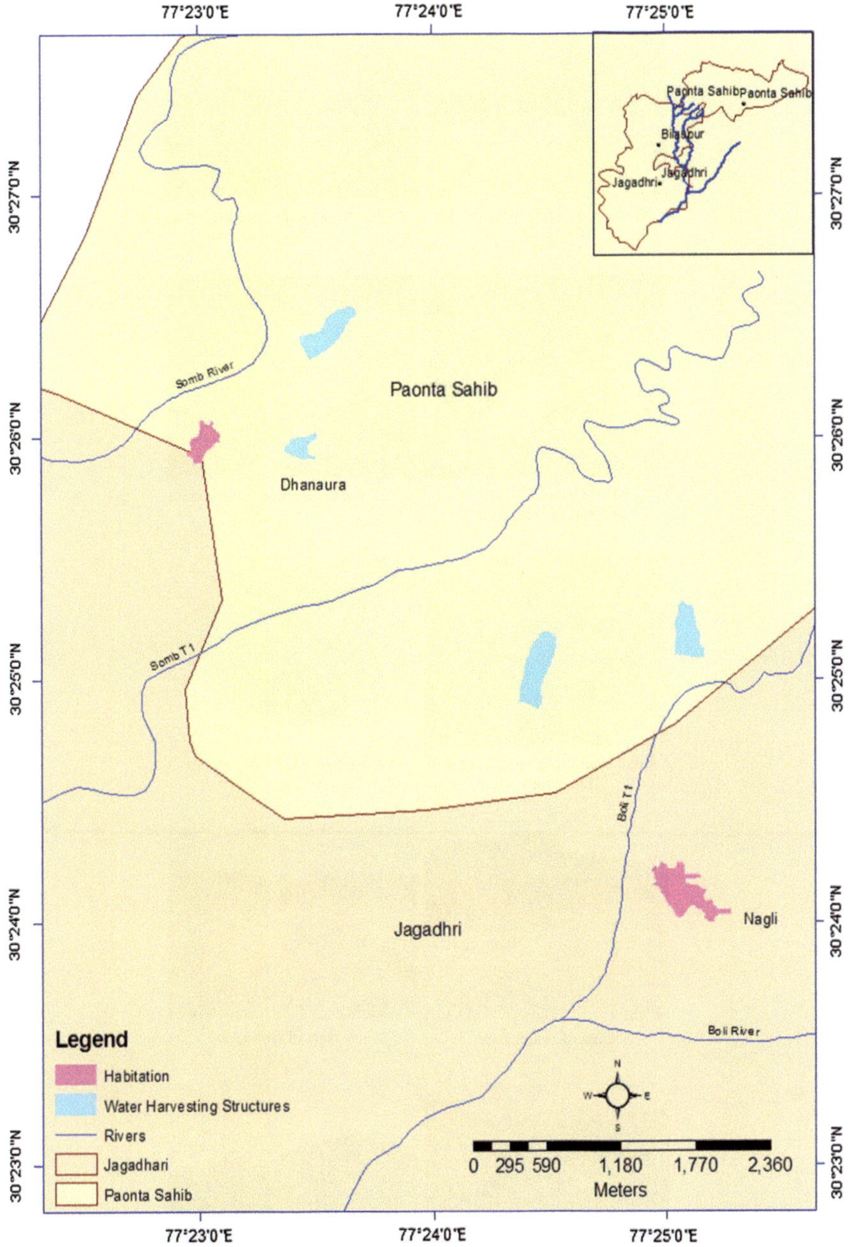

Fig. 4.1 Location map of Dhanaura and Nagli village (including Somb and Boli river tributaries of river Yamuna)

Water harvesting structure in Dhanaura

Waste Weir for Water harvesting structure

Check Dam for prevention of soil erosion in Dhanaura

Solid waste disposal in Dhanaura

A small household biogas plant in Dhanaura

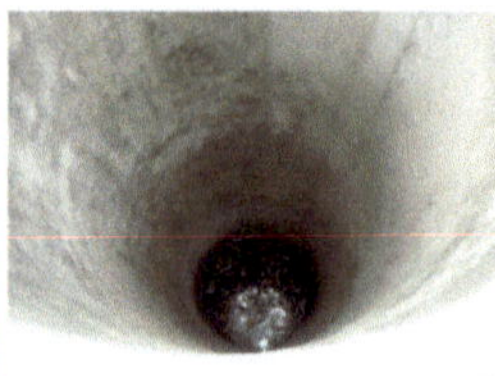

Abandoned well in Dhanaura

A view of Somb river

A view of Boli river

Water harvesting site in Nagli village

Natural water storage site being developed for recreation in Nagli

Fig. 4.2 Different view of Dhanaura and Nagli village (including Somb and Boli river tributaries of river Yamuna)

contributes 81 % of the annual rainfall in the district, commences late in June and continues up to about the middle of September. The average annual rainfall in the district is 1,116 mm.

4.2.2 Soil, Land Use and Land Cover

The soils in these villages are generally good alluvial loam, though not so rich. Soil has low nitrogen (OC, 0–0.4 %), low phosphorous (as P_2O_5, 50–10 kg/ac) and medium Potash (K_2O, 55–135) [Department of Agriculture, Haryana (2007–2008)]. Both the intervened villages require further determination and capacity building of the institutions developed to work forward for integrated nutrient and pest management (INPM). Land use in the area can be broadly classified as agriculture, hilly forests (protected), and water bodies.

4.2.3 Biodiversity

Sal (*Shorea robusta*), Semul (*Salmelia malberica*), Amaltas (*Cassia fistula*), Bahera (*Terminalia bellerica*), Khair (*Acacia catechu*), Shisham (*Dalbergia sissoo*), Chhal (*Anogiessus latifolia*) and Peepal (*Ficus reliogiosa*) are the main trees in the nearby forest of Dhanaura and Nagli. Forests are of northern dry mixed deciduous type. There are a number of miscellaneous hardwood species found in the forests. Other species include Jhingan (*Lannea coromendelica*), *Reru* (*Acacia leucophloea*), *Phulah* (*Acacia modesta*), and Bamboo (*Dendrocalamus strictus*). Some *Chir* (*Pinus* spp.) forests are found along the ridges in hills. However, bamboo is mostly localized on northern hills and in the gullies.

The main shrubs include *Khuri* (*Nyctanthesarbor-tristis*), *Garna* (*Carissa spinarum*), *Basa* (*Adhatoda vasica*), *Gandhera* (*Murraya koenigii*), *Mandhar* (*Dodonia viscose*) and *Phullakri* (*Lantana camara*). The important grasses are: *Dhaulu* (*Chrysopogon montanus*), *Sariali* (*Heteropogon contortus*) and *Bhabbar* (*Eulaliopsis binata*), *Chrysopogon fulvus*, and *Pennisetum purpurium*.

Forests of the area are known to be the source of a large number of valuable medicinal plants. Few of the medicinal plants such as *Agave*, *Aloe*, *Datura stramonium* are of great demand. Some of the reported threatened medicinal plants of the region are: *Aegle marmelos*, *Oroxylum indicum*, *Symplocos racemosa*, *Gloriosa superba*, *Celastrus paniculatus*, *Pueraria tuberose*.

The Non Timber Forest Products (NTFP) like gums, resins, essential oils, tans, dyes, fibres, bamboos, medicinal plants, etc. have great potential in national and international markets. There are a number of species in these forests yielding NTFPs, which need to be identified, propagated, planted and exploited in a systematic manner.

National Medicinal Plants Board, Department of AYUSH has helped various villages of block Chhachrauli, Yamuna Nagar for cropping of medicinal and aromatic plants. These plants are *Aloevera, SafedMusali, Amla, Beal, Ashok, Sarpgandha* and *Mulethi*. Such an activity needs initiation in this village also for revenue generation.

Fauna: Wild boars, sambhar, dear, black buck, chital, jackal, fox, bear, tiger, leopard, monkey, apes, blue bulls, rabbit, and porcupine are main animal species in the forest. Forests have various bird species such as peacock, titar, bater, cuckoos, drongo, hornbills, blue peafowl and partridges, and among replites, varanus, lizards, chameleon, cobra, python and green wipe snake are common.

4.2.4 Plantation

Both the villages claim that their share of village area under forest has been taken by the Forest Department. Now villages have no forest land. Forest Department does not allow villagers to enter in the forest. Villagers prefer agriculture than plantation or farm forestry. They have planted *Safeda* and *Poplar* on the farm boundary for revenue generation. Among fruit plants mango, guava and banana are common. The area is mountainous and as a follow up of forest policy they can't cut trees in their own field except *Safeda* and *Poplar.*

Villagers face problem for fuel wood. Native shrub species may be promoted for fuel wood. The exotic plant species (poplar) has became more common than the indigenous one due to local demand of paper industries. Commercialization has led to the diversion of good agricultural lands to forestry uses. Thus, the social forestry program under the process of implementation has suffered serious distortion and needs to be reoriented so as to be more meaningful. *Safeda* and *Poplar* requires more water and therefore cannot be commercially planted. Indigenous shrub species will be suitable for that.

4.2.5 Common Crops

Common cultivated crops in Dhanaura and Nagli are: *Bajra*: *Pennisetum glaucum, Paddy*: *Oryza sativa, Wheat*: *Triticum aestivum, Jowar*: *Sorghum, Millet*: *Urochloa ramose, Maize*: *Zea mays, Brinjal*: *Solanum melongena, Potato*: *Solanum tuberosum, Tomato*: *Lycopersicon esculentum, Onion*: *Allium cepa, Garlic*: *Allium sativum, Ginger*: *Zingiber officinale, Sarson*: *Brassica campestris, Desi-chana*: *Cicer arietinum (chickpea), Masur*: *Lens culinaris (lentil), Tur (Cajanus cajan, pigeon npea, arhar), Urd*: *Vigna mungo (black gram), Mung*: *Vigna radiate (green gram), Sem*: *Lablab purpureus (lablab bean, vaal), Baqla (Vicia faba, faba bean) grows wild, Peas*: *Pisum sativum, Alsi*: *Linum usitatissimum (Linseed -Oil), Til*: *Sesamum indicum, Ground nut*: *Arachis hypogaea, Sarso*: *Brassica compestris* (*mustard*), *Chillies*: *Capsicum cultivars*, *Turmeric*: *Curcuma longa*, *Sunn hemp*: (*Crotalaria juncea*, *Kapas-cotton*: *Gossypium herbacium*, *Sugarcane*: *Saccharum officinarum*, *Barley*: *Hordeum vulgare*, *Bhang*: *Cannabis sativa*, *Mango Mangifera indica*, *Banana*: *Musa acuminata*, *Guava*: *Psidium guajava*, *Safeda*: *Eucalyptus tereticornis*, *Poplar*: *Populus euphratica*, *Berseem*: *Trifolium alexandrium(fodder)*, *Oat*: *Avena sativa (fodder)*, *Oat*: *Medicago sativa (fodder)*, *Lucerne*: *Alfaalfa (fodder)*, *Chari*: *Sorghum bicolor and Sorghum cultivars (fodder)*, *Lobhia* (*Vigna unguiculata*, cowpea, C*howli*) (also fodder), etc.

4.3 Inference from Primary Survey

Village elders, farmers, government persons, and women were interviewed/consulted separately. Group discussions were also conducted to know the mass awareness regarding the natural resources and related problems. In both the villages interventions were done for soil and water conservation, and forest management. Inferences were drawn following the category of interventions and watershed development projects, environmental status, socio-economic status, agriculture and irrigation, livestock, amenities, institutions, etc.

4.3.1 Integrated Watershed Development Project

Integrated Watershed Development Project (IWDP) of Haryana favored water harvesting for perennial sources of water and supplements improved agriculture, land management with the help of local institutions. The main objective of the IWDP is to restore the productive potential of the project area by developing soil–water plant resources of the watershed to rectify the man made ecological imbalance by conserving the natural resources on arable and non-arable lands. The emphasis is to conserve soil and moisture in situ through integrated approach with bottom up planning and top to bottom treatment. IWDP also worked for capacity building for local institutions for maintenance of intervention. Villagers were trained on principles and practices on participatory natural resource management as well as integrated watershed management.

Forest Department, Haryana and TERI by funding of Ford Foundation in Year 1990–1992 worked for Joint Forest Management (JFM). About 50 Hill Resource Management Societies (HRMS) have been formed in Shiwalik foot hills near Yamuna Nagar. HRMS worked in association with Forest Department. Under this scheme water harvesting structure and check dams were constructed. Watershed development and forestry development by the Forest Department and under IWDP Programme of Department of Agriculture, Haryana by support of Ministry of Land Resources (World Bank funded) has improved the water resources of area.

Three water harvesting structure have been constructed in village Dhanaura through public participation and government support (under IWDP and Forest department). Two small check dams/silt detention tank to avoid soil erosion and siltation in these structures have also been constructed. Villagers have formed a society, Van Sampada Sanrakshan Samiti (VSSS) for the management and distribution of water from these water structures for irrigation. They follow the water distribution rule with charge of Rs 10/- per hour irrigation. The *Warbandi* (day wise allocation) system for distribution of water is adopted. The revenue generated is used for system maintenance. In pre-monsoon, dredging is done for maintaining the capacity of the structure. VSSS also works for dredging/desilting of the structures through public participation.

These structures have improved ground water level. Water table nearby the structure has been lifted up to 4.5–6.0 m, though the average water table in the village is at 15.0–18.0 m. A good storage of water was recorded at the time of visit (August 2009) in the structures. Villagers informed that water remains in the structures till winter. In Dhanaura, the stored water is generally used for irrigation.

Nagli has three (2 government and 1 private) water harvesting structures and 4 soil detention check dams. One was constructed in 1995–1996 under IWDP. In the year 2005–2006, Forest Department with JBIC funding constructed water harvesting structure in Nagli through public participation. One private water harvesting structure is being developed for recreation purpose near mosque. This is really good start where we find that the development done has been finally adopted by the inhabitants and they are themselves taking initiatives. Such cases need to be promoted in hilly undulated terrains for water conservation.

Government agencies had installed two filters in river Somb to provide water to Dhanaura. These filters worked effectively for 2–3 years and later on they are either buried in mud or wiped away by water in river. Villagers informed that filter was very useful however, it did not worked long. At the time of visit, stone mining was being done from the river bed for construction work.

4.3.2 Environmental Status

Drinking Water: Government tube well is a major source for drinking water; each house has water connection through pipes. Water supply is affected due to frequent electricity failure. Earlier, villagers used to take drinking water from the dug wells. But now as there is piped water supply these dug well are not in use and therefore abandoned.

Water Table: Presently, water table in the dug wells of Dhanaura is at 24.0–26.0 m, however, it was 15.0–18.0 m till 2001–2002. The declining of water table in the wells is the main reason of abandonment at Dhanaura.

A similar picture was observed in Nagli. The water table is very deep up to 21.0–24.0 m deep. Presently, 4–5 dug well are in use out of 12 dug wells. For supplementing the water demand, submersible pumps are being used to extract the

Table 4.1 Average water table in Dhanaura and Nagli

Village	Water table (m)					
	1995–1996		2001–2002		2009 (low rainfall)	
	pre- monsoon	post monsoon	pre- monsoon	post monsoon	pre- monsoon	august
Dhanaura	16.8	15.2	19.5	18.3	26.5	25.3
Nagli	19.8	18.3	21.9	19.8	26.2	23.5

Source Primary survey (discussion with villagers)

groundwater, for which pipe casing up to 100–180 m is installed. Seasonal water table fluctuation in the villages can be seen from Table 4.1.

In general, dug well are not used in both the villages for drinking water. In both the villages, an overall declination in the water table can be seen due to over exploitation of ground water, though the recharge has increased through the water harvesting structures. Due to deep water level, centrifugal pumps cannot be used (completely stopped from last 6 to 7 years) and has been replaced with the submersible pumps. This feature in the villages leads to over dependencies on the electricity supply for water. Submersible pumps are being used for lifting the ground water for irrigation as well as for drinking. Water quality of tubewell is adequate for drinking water supply.

Sanitation and Hygiene: Only 10 household have toilet facility in Dhanaura. Government helped the village by making 4 toilets in village for scheduled caste category (SC), but presently defunct due to lack of usage and water supply. Condition of Nagli village is even more worst as only 5–6 household has toilet facility. It means sanitation system is largely depends upon the water supply and thus the electricity. Therefore, in the rural areas, it will be better to promote indigenous sewerage and sanitation system rather than the advanced one considering the large seasonality in the system.

Waste Disposal: Generally open field spaces are being used for wastewater disposal. Solid waste is disposed into a pit which is used as organic manure.

Disease: Malaria is common disease found in village. Impoundment of wastewater near by the habitation multiplies the cases of malaria in the villages.

Fertilizer Application: Urea, di-ammonium phosphate (DAP), potash are the main fertilizers used in fields for supplementing the Nitrogen (N), Phosphorous (P) and Potassium (K). Green manure is also used, but low livestock population has increased dependency on chemical fertilizer. Soil of this area is not rich in nutrients therefore use of leguminous crops and organic manure need to be encouraged. This will also help in better management of crop water demand.

Pesticides Application: Endosulphan is common in use, depending on crops. No integrated pest management or traditional methodology is used. Local institution should take the responsibility of integrated pest management (IPM) with the help of agriculture department. Public partnership in IPM is lacking because government has not taken any initiatives.

Problematic Wild Animals for Agriculture: Neel gai/blue bull (*Boselaphus tragocamelus*) and wild boar (*Sus scrofa*) damages crops. Many field are abandoned due to these, villagers have planted Safeda to many field to avoid further damage in both the villages. Safeda provides revenue as well as fuel wood and timber.

Fuel: Wood and cow dung cake are used as fuel. About 6 household have micro biogas plant and 15–20 household has Liquid Petroleum Gas (LPG) connection in Dhanaura. However, in Nagli, villagers only 8–10 families are using LPG. Forest developed under JFM are used under strict rules to collect tree lopping and dry twigs for 6–9 months at the rate of Rs 7 per month for felling of 100 bamboos. Villagers store the dry fuel wood from forest for house hold usage.

Villagers have planted safeda and poplar which have commercial importance. There is need of plantation of species such as *Casuarina equisetifolia*, *Dalbergia sissoo*, *Acacia nilotica*, *Leucaena leucocephala*, *Sesbania gradiflora*, *Gmelina arborea*, *Callicarpa macrophylla*, *Carissa spinarum*, *Clerodendron viscosum*, *Colebrookia oppositifolia*, *Murraya* sp., *Woodfordia* sp. *Zizyphus* sp for wooden fuel and fodder. In spite of fuel fulfilment, these plant species are useful for erosion control.

4.3.3 Socio-Economic Status

Habitation: *Jat*, *Gurjar*, *Bania*, *Dhobi*, *Teli*, *Harijan* are main residents in Dhanaura. Among all, Jat population is prominent. Dhanaura village has 100 households with population of nearly 600. On the other hand, Nagli is a minority village of 500 households with 5,000 people. Its residents are muslims (prominently) followed by Jat, dhobi and harijan.

Land Holding: Land holdings in both the villages falls within small to medium. In Dhanaura, 15–20 household are under marginal farmers (<0.5 acre), 10–15 are landless, and rest of the household belongs to small (5–7 acre) to medium (7–20 acre) land holding. Land holding in the Nagli village is generally small to medium scale. Only 4–5 farmers belong to the category of large farmers (i.e. land holding >20 acres).

Occupation: Agriculture is the main sources of occupation for the villagers. Small land holding and landless household are shepherds and keep sheep and goat. A few people work in Jagadhari. Village people do not have specific professional skill to get off-season employment. Small scale industry and training programme for women like weaving, embroidery, basket making etc. need to be encouraged for revenue generation. High unemployment was noticed in villages, a large number of productive human resources can be used as pool for seasonal employment. Mahatma Gandhi National Rural Employment Guarantee Programme (MNREGA) has also not assisted these villages. It was also noted that the Villagers have the common consensus for not sending their children for secondary education because (a) colleges/vocational training institute are not available near by the village, and (b) lack of employment opportunity.

This area has good scope of growing medicinal and aromatic plants such as *Aloe vera*, Safed Musali, Amla, Beal, Ashok, Sarpgandha and Mulethi for revenue generation, and need to be encouraged. HRMS and VSSS may work for this also. Though rainwater harvesting has been done but still half of the culturable land left unsown due to lack of water as reported by the villagers. That area can be utilized for such farming as these also require less water and generate revenue. Therefore, rainwater harvesting need to be further emphasized in these area to augment 100 % irrigation through MNREGA.

Migration: Migration is not evident in these villages due to lack of skill and education. People fulfil their need from agriculture. Few people works at Jagadhari. During summer, Shephards move to Haridwar and Saharanpur, but this is not a regular practice. However, people during non-farm season need certain employment but lack of skill and unavailability of employment for skilled/ unskilled in the nearby area reduce their chance. Local institutes should take initiative to provide vocational training to the villagers.

Status of women: Women in Dhanaura are aware for the water resources as well as for environment. Families send their daughters to school, and many girls study up to high school (Metric) in Dhanaura, whereas, in Nagli village importance on girl's education has not been given. Very few families send their girls to Madarsa. Girl's education in Nagli needs to be promoted. Water harvesting structures have helped women financially because of secured source of irrigation and it has increased the irrigation. Fuel wood availability is fulfilled by dry woods and twigs of poplar and *safeda* which have been grown in villages. Earlier, they used to go 3–4 km for wood collection. Drinking water availability through piped water from tube well has provided good time to women for in-house works otherwise they used to bring water from open dug wells.

Women in both the villages are very much interested in-house employment such as embroidery, basket making etc. however, there is no such encouragement by government or local institutions. A proper attention need to be given in this regard, which help in revenue generation. Women are aware of the excess use of water in their current cropping pattern. However, they say that if they have alternative sources such as pasture, seasonal employment etc. then they can shift to their previous cropping pattern.

4.3.4 Agriculture and Irrigation

Agriculture: Dhanaura and Nagli both have two crop seasons, *Kharif* and *Rabi*. Sugarcane, wheat and rice are the main crops in Dhanaura. Berseem is grown as a fodder crop for livestock. However, a variety of plants grown in the field earlier have been either cultivated on small scale or stopped. Diversity for cultivated plants has reduced and dependency on few crops has increased, because of availability and exploitation of ground water for irrigation.

Table 4.2 Average crop production in Dhanaura and Nagli (Kg/ha)

S.No.	Crop	Dhanaura		Nagli		Yamuna-Nagar[a]
		Before intervention	After intervention	Before intervention	After intervention	
1	Wheat	800	1,000	850	1,000	3,668
2	Rice	1,550	2,000	1,600	2,000	3,020
3	Sugarcane	1,400	3,700	1,500	3,500	5,944
4	Maize	1,250	1,550	1,200	1,600	1,779
5	Urad	130	160	150	200	206
6	Til	80	150	80	150	200
9	Masoor	450	500	600	700	836
10	Gram	100	150	90	200	300
11	Jowar	60	80	70	80	100
13	Bajra	550	700	500	700	891
14	Ground nut	260	300	350	400	755
15	Potato	500	700	560	700	1,051

Source [a] secondary data

Nagli has variety of crops such as *Wheat, Chana, Masoor, Urad, Rice, Maize* and *Bajra*. Nagli village has not lost the crop diversity as compared to Dhanaura, which reduced the crop risks under adverse climatic conditions. The village institutions have also created zeal for conservation of water and cropping fodder trees.

Crop production: Villagers informed that water harvesting and soil conservation measures helped in the irrigation but productivity has not significantly improved. This might be due to soil fertility and dependency on chemical fertilizers and change in cropping pattern. The average crop production in the villages are presented in Table 4.2.

Primary survey confirms the overall increase in the production; it may be due to the availability of water, better usage of seeds, fertilizers and pesticides. However, the production is far below from the average statistic of the Yamuna Nagar (Table 4.2). The comparison of Tables 4.1 and 4.2 clearly indicates that the ground water exploitation has been significantly increased in the area and thus the groundwater table is declined because of increased acreage of the high water demanding crops like paddy and sugarcane.

Considering this, for revenue generation certain crop of low water requirement needs alleviation. Rice, Wheat and Sugarcane are main crops in Dhanaura while Nagli depends mainly on Rice, Wheat, Sugarcane and mixed crop of maize and *Urad*. The soil quality and topography do not support dependency on only rice, wheat and sugarcane. Other low water consuming crops, horticulture, leguminous crops could be beneficial for survival and revenue generation. Based on the secondary data (1991–2007), the seasonal cultivated areas under different crops are shown in Figs. 4.3, 4.4, 4.5, and 4.6, which indicate that the crop diversity is decreasing in these villages.

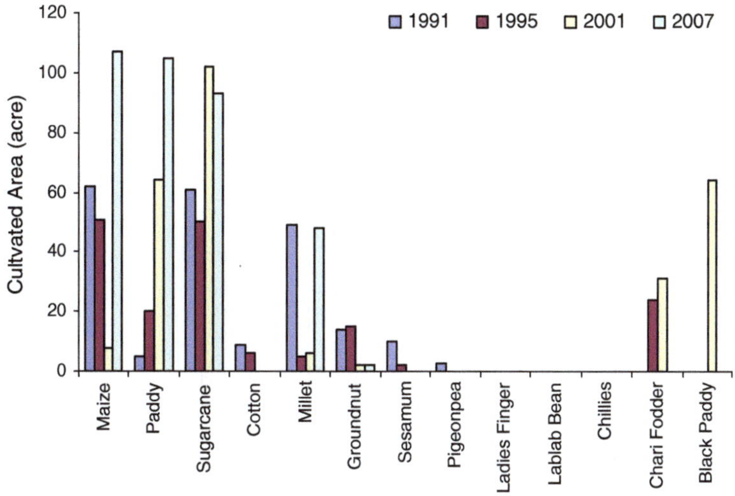

Fig. 4.3 Cultivated areas in Kharif season in Dhanaura

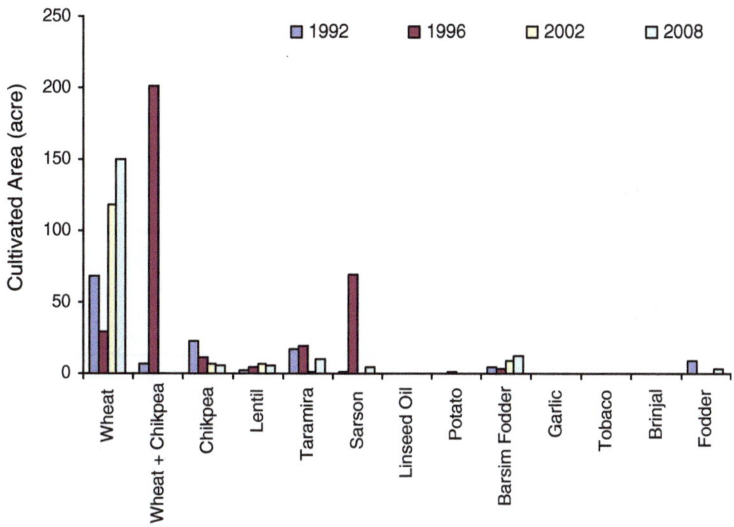

Fig. 4.4 Cultivated areas in Rabi season in Dhanaura

4.4 Summary and Conclusions

Based on the above case studies of two villages in the Somb and Boli catchments of the Yamuna river basin, the summary and recommendations have been made and scaled up to the hilly catchments of Shivalik range (Table 4.3).

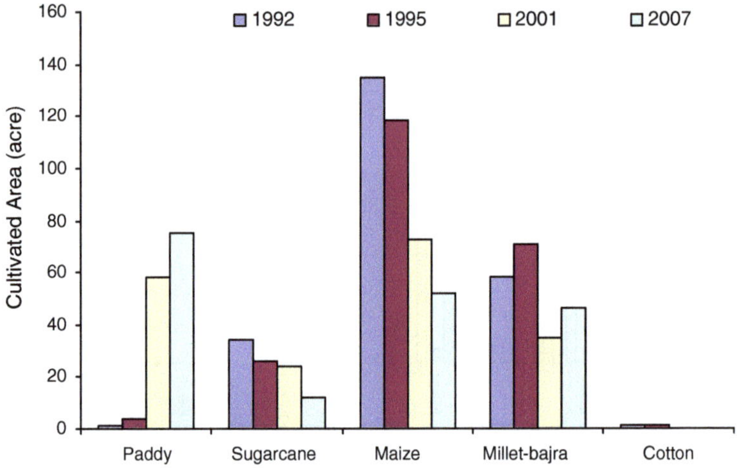

Fig. 4.5 Cultivated areas in Kharif season in Nagli

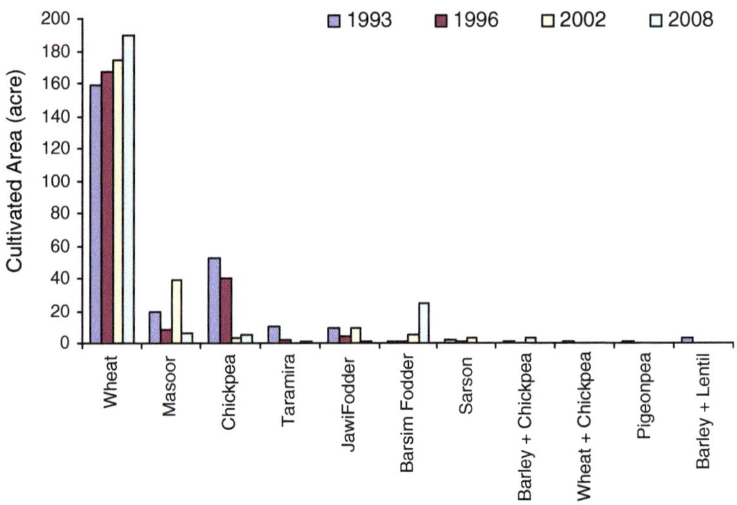

Fig. 4.6 Cultivated areas in Rabi season in Nagli

It may be concluded that public participation in any benefiting development programmes is high. However, it is also appeared that public has not taken any independent initiative in the watershed planning and management without the

Table 4.3 Summarized status and possible recommendation of the issues

S. No.	Issue	Status	Concluding Remarks/Recommendation
1	Education	Inadequate amenities in primary education	Schools as well as vocational training centres are required
2	Drinking water	Adequate	Abandoned dug wells may be cleaned for usage
			Secure power supply is required to maintain the water supply
		Ground water (GW) level is lowering	GW utilization at sustainable basis should take place in that area. High water consuming crops like Paddy and sugarcane has increased the groundwater exploitations in the area Therefore, for sustainable planning of the resources, agriculture sector should be given more importance and allocation of cultivable land should be planned based on the climate, water and land potential
			For increasing GW recharge a simple technique that could be developed by the villagers themselves like diverting natural streams to a recharging wells or injecting wells through piped or open ditch. For that masonry or a stone structure can to be constructed on the river to divert water to the recharge well, which leads to GW recharge. Dhanaura may use those 3 abandoned recharge wells and make them operative by some community involvement rather than depending on government
			While, in Nagli same could be done to immensely reduce the depletion of GW
3	Sanitation and hygiene	Open defecation	Community toilet may be proposed
		Waste disposal problem	Disposal of domestic wastewater is a major problem in these area or region. Soak pits shall be usable for these areas, which are simple and cheap drainage structures facilitating the disposal and filtration of wastewater. The cost per unit of soak pit is very less (Rs 300–400/unit) and can be constructed by help of local institutions through public participation
4	Electricity	Frequent power failure (6–7 hr supply). Many times 2–3 days of power cut	Daily power supply should be maintained so that water supply can be managed. Alternate energy sources especially the solar energy need to be promoted by the government initiative. It will not only maintain the continuous water supply, but also helps in increasing the irrigation efficiency using the efficient irrigation methods like sprinkler and drip

(continued)

Table 4.3 (continued)

S. No.	Issue	Status	Concluding Remarks/Recommendation
5	Land resources	Severe soil erosion on slopes and along streams	Conservation measure required at many point in Dhanaura as well as in Nagli
		De-silting	The sedimentation decreases the effective storage capacity of check dams and the percolation of water. Invariably removal of silt from the storage basin is laborious and expensive. Some curative measure needs to be taken to reduce silt load on to check dams. Gully plugs could be constructed on upstream of the check dam, which will slow down water and will help in GW recharge along with reducing silt load into the check dams
			For sustainable erosion control in the area, catchment area treatment should be done
6	Forest	Shrinking forests Lack of water and food led the wild animals destroy surrounding croplands	Forest is shrinking in space and density. Villages should have forest and pasture for livestock grazing and other by products for revenue generation. Protected area should be developed for limited use so that the regeneration capacity of the forests is achieved. This will also help in checking encroaching forest of Forest Department
			As the forests do not provide the requirement of wild animals they destroy cropland of the adjacent villages. Forests should be barricaded through vegetation as well as by wires to check passage. Plantation of fodder and fruit trees will attract neelgai and monkeys to stay in forest only
7	Water Resources	High surface runoff Very few water harvesting structures	As the area is hilly the runoff is very high during rains. This leads to soil erosion and loss of water from the catchment
			Additional water harvesting structures and soil conservation structures are required
8	Agriculture	Low crop yields Increasing cultivation cost Low crop diversity	Soil conservation measure and usage of organic manure should be increased
			Crops with low water demand should be promoted. This will reduce the cost of cultivation and over exploitation than GW recharge
			Crop diversity should be enhanced for secure economy of village as well as better crop- water management

(continued)

Table 4.3 (continued)

S. No.	Issue	Status	Concluding Remarks/Recommendation
9	Irrigation	Tube well Water harvesting structures	Tube well with submersible pump system is used for irrigation Water harvesting structures are also used for irrigation water Water quality is all right, however extraction of water is very high
10	Livestock	Lack of fodder management	Livestock are major part of rural economy. Integrated water resource management addresses the whole requirement of a watershed at a small as well basin on large scale. Plantation and regeneration of fodder tree needs to be prioritized Protected pastures and farm forestry land may be developed as these villages have tremendous scope for these
11	Non farm sector	Not scaled up	Non-farm activities may include basket making, embroidery etc. Women may be encouraged to work for these for revenue generation
12	Economic security	No stabilized income Risk during aberrant weather	Crop diversity, pasture, farm forestry and non- farm activities should be implemented to stabilize income. As the area is at high risk of erosion, damage of crops by wild animals, some secure income space is urgently required. Villagers are also very enthusiastic to these initiatives
13	Institutions	Van Sampada Sanrakshan Samiti and hill resource management society	Van Sampada Sanrakshan Samiti and Hill Resource Management Society both have limited their scope for construction of structure Villagers may be trained by these for better livestock and farming and non-farm activities for revenue generation Non timber forest products are very important for revenue generation these may be supported by the institutions for management Also the power of these societies is limited with few hands and all the villagers are not benefited by these directly
14	Landscape	Landscape transformation	Forests are declined in size when new roads or any development is initiated. Institutions can work effectively for sustaining forests in association with development. Awareness about the changing effects of humans on landscapes with shrinking forests, pressure on agriculture and irrigation can serve as a useful tool in formulating regionally appropriate policies

government initiatives because of unavailability of funds, knowledge, and insufficient institutional set-up. It is also true that the government and non-government agency cannot successfully implement any program of IWRM approach without the consensus of local public and their partnership.

References

CGWB (2010) Groundwater status of India. Central Ground Water Board, Faridabad, India

Rai RK, Upadhyay A, Ojha CSP, Singh VP (2011) The Yamuna river basin: water resources and environment. Water Science and Technology, vol 66. Springer, The Netherlands

Chapter 5
Case Study: Middle Alluvial Plains

Abstract This chapter presents the case studies in the catchment of middle alluvial plains. In this catchment, both water quantity and quality has the major challenge. These problems are not inherent rather caused by the upper catchments up to Okhla barrage. Because of quality, even drinking water is not sufficiently available to the mankind. This catchment requires an immediate attention for IWRM.

The middle plains of the Yamuna river basin is referred as plain along the river reach from Okhla barrage to the Etawah before the confluence with river Chambal. This reach is largely vulnerable to the water quality deterioration because of disposal of partially treated or untreated wastewater into the river. In this alluvial plain at the right bank of the river Yamuna, Agra canal supplies irrigation water to its command area up to Agra. The Agra canal water is getting water from the Okhla barrage and Hindon River through Hindon-cut, and has very bad water quality because of wastewater effluents from the Delhi NCR (Rai et al. 2011). This water is not adequate for irrigation as per the standard and affecting the land and groundwater quality of the alluvial plains. Because of this concerned water quality issue in the region, public participation campaign has been conducted. For which, two villages Ajnauthi (of Undi Panchayat) and Badawali (of Badawali Panchayat) of Chhata block (near Kosi Kalan, an industrial town) of Mathura near by the middle alluvial plain were identified (Fig. 5.1). These villages of course using Agra canal water (untreated/partially treated wastewater) for irrigation (Fig. 5.2). When excess nutrient available in the wastewater, the groundwater quality will deteriorate and may cause health implication when used for drinking. An insight into the status of water management programmes through IWRM principles and their impact on the changes in land use, agriculture, livelihood, water quality in these villages has been investigated.

A. Upadhyay and R. K. Rai, *Water Management and Public Participation*,
SpringerBriefs in Earth Sciences, DOI: 10.1007/978-94-007-5709-7_5,

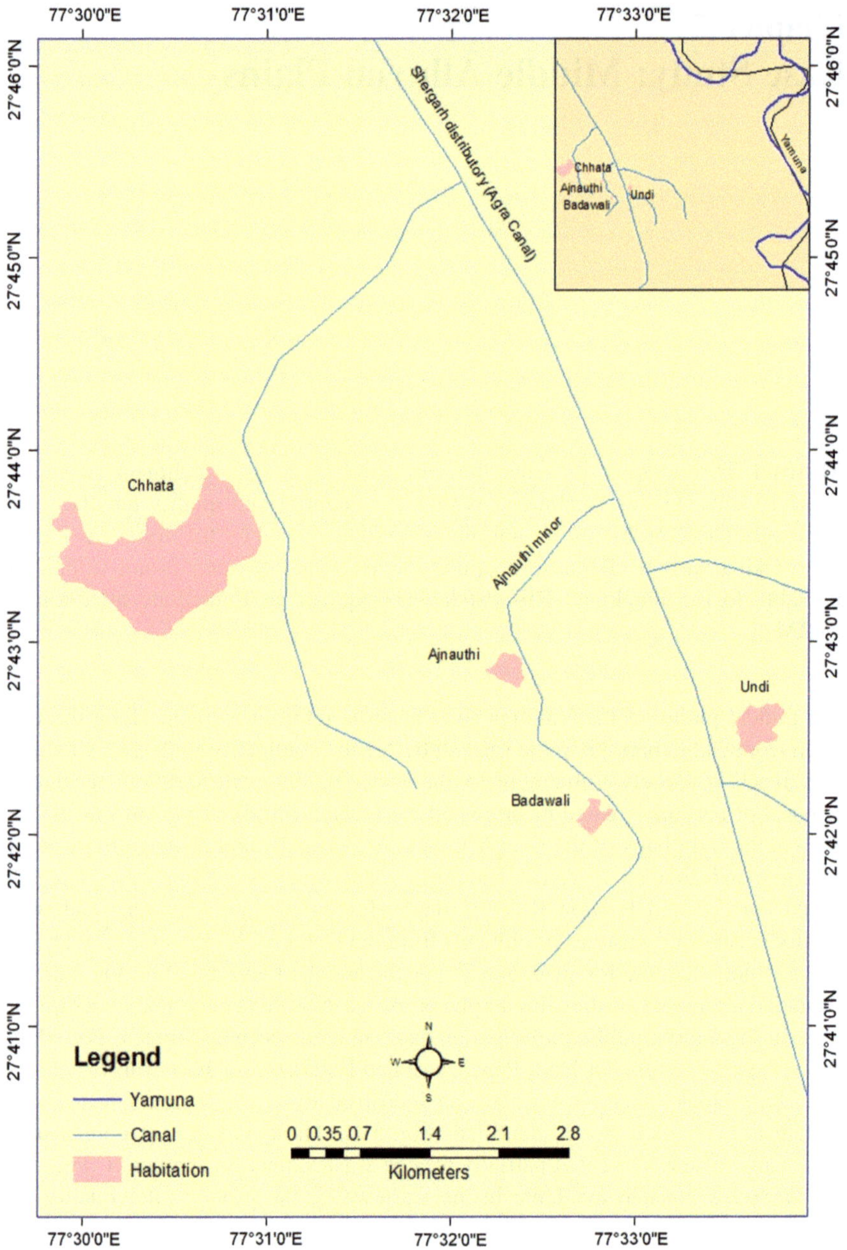

Fig. 5.1 Location map of Ajnauti and Badwali village of Mathura

Fig. 5.2 Status of Agra canal water

5.1 Geo-Physical Characteristics

In the physical characteristics, climate, soil, landuse and land cover, and biodiversity has been discussed.

5.1.1 Climate and Rainfall

The climate of area can be categorized as semi-arid climate that experiences extreme summer and winter temperatures. The average rainfall is of the order of 593 mm, which is mostly confined during the months of July–September. The summer temperatures range between 22 and 45 °C, whereas winter temperature varies between 14 and 32 °C. Geographically, the area lies in the rolling fertile plains of north central India.

5.1.2 Soil, Landuse and Land Cover

In this segment, topography is almost flat. Mathura district falls under south western semi-arid agro-climatic zone of Uttar Pradesh and mainly termed as western plain region. In general, soil in the area belongs to sandy to sandy loam and loamy texture. Land use of the area can be broadly classified as agriculture, pasture, habitation and waste land. There is no forest cover experienced in the area, though the forest area is available. Approximately, 60–65 % areas in the villages are used for agriculture. There is no distinct pasture land is available for grazing. Forest Department has not worked as such for the village. Villagers informed that their share of land of forest has been taken by Forest Department, and they have not developed that land yet. Even if they develop forest or pasture that will be of no use to village directly as they will not get access to it.

Paddy, wheat and sugarcane are the major crops cultivated in the area. Other than the major crops, Jowar, Bajra and Sarson, Urad, Til, Maize, Cotton, Barley, etc. are also cultivated in the village.

5.1.3 Biodiversity

Neem, Pipal, Amaltas, Pakar, Sheesham, Mango, Babool, Dhaincha, Kikar, Vilayati Babool and Ber are the main trees found in the village. Based on the villager's information, wolf, jackal, blue bull are the main mammal species, and crane, woodpecker, crow, maina and fakhta are the common birds species found in the area.

5.2 Inferences from Primary Survey

These villages have no intervention like rainwater harvesting or other conservation efforts. Primary inspection based on the greeneries, it was appeared that groundwater table is quite high in spite of its quality.

These villages have the water user association (WUA) created by Uttar Pradesh Irrigation Department. They receive irrigation water from Ajnauthi minor of Shergarh distributary. However, the WUA as such is not functional and do not supported by the existing Krishi Vigyan Kendra (KVK) of Indian Council of Agricultural Research (ICAR), Ministry of Agriculture (MoAg) for better cropping and integrated nutrient and pesticide management. It reveal a large institutional gap who responsible for common goal of agricultural productivity and food security (i.e. Agricultural department and irrigation department). Therefore, for successful implementation of IWRM, institutional set-up comprising of various knowledge-base and stakeholder involvement with public participation is mandatory.

5.2.1 Environmental Status

5.2.1.1 Drinking Water

Ajnauthi and Badawali has problem of saline water in tube well as well as in dug well. In Ajnauthi only one dug well is having fresh/sweet water and inhabitants are dependent on that well (Fig. 5.3). All the dug well are abandoned due to saline

Fig. 5.3 Water problem in Ajnauti

water. Badawali has a little better situation with respect to water quality in dug well.

Deep tube well in both the villages provides saline water. Sweet water is available up to 12.0–15.0 m. However, water at this level is not sumptuous. Beyond this depth, saline water problem is encountered. Villagers are forced to irrigate with saline water of deep aquifer.

5.2.1.2 Groundwater

Villagers take drinking water from the dug well. Jal Nigam was working for piped water supply through a tube well to Ajnauthi but that work has been postponed after site selection and initial excavation work may be because of salinity problem. Villagers had complained about the drinking water problem at block level, however, no action has been taken. Groundwater in the village is saline and yellowish in color. It leaves white salt after evaporation, which can be seen on the walls of the houses and open fields. In both the villages, water table fluctuations during the pre and post monsoon are very high (Table 5.1).

Water level in the western zone of Uttar Pradesh which comprises 252 development blocks has 70 dark and 64 grey zones (CGWB 2010). Salinity problem in this area including these two villages has also been reported in the Integrated Management Information System of Department of Drinking Water Supply (as on April 2009). Secondary data of Central Ground Water Board for Mathura shows problem of water quality in the ground water (Salinity as EC > 3,000 μS/cm at 25 °C, Fluoride > 1.5 mg/l, Chloride > 1,000 mg/l, Nitrate > 45 mg/l, Arsenic > 0.05 mg/l). Fluoride problem has also been reported in Chhata block (CGWB 2010).

Water availability in arid regions is both sporadic and highly variable in quantity. If the water quality shows large variations of salinity and concentration of other chemical constituents with depth and time span, it has considerable effect

Table 5.1 Pre and post monsoon water level in Dug well (m bgl)

Village	1995		2008	
	Pre monsoon	Post monsoon	Pre monsoon	Post monsoon
Ajnauthi	6.0	1.5	9.0	1.5
Badawali	7.0	1.5	10.5	1.8
Chhata[a]	–	–	9.0	1.5

Source Primary survey; and [a] secondary data from CGWB

on the entire hydro-geological set-up of the area. The deep aquifers that supply water to bore wells in the alluvial plain of the Mathura region have higher salinity than those of the dug wells from the shallow aquifers. The excessive drilling of tube wells and high yield tube wells are resulting in deterioration of water quality of the shallow aquifers.

5.2.1.3 Sanitation and Hygiene

In both the villages no construction has been done for wastewater disposal. There are no proper wastewater drains, and wastewater percolate to the low land area near village, which cause groundwater quality deterioration and therefore health implications. Due to this reason stomach problem in the villages are noted. Open defecation is practiced as no toilet facility either community or private is available.

5.2.1.4 Waste Disposal

Open dumping is only way practiced for disposing waste. At few place in the village, waste is disposed in low land area for degradation. However, barricading of disposal site was absent causing spreading of wastes over the streets. There are various government schemes for waste disposal and sanitation have been launched, but any initiatives have not been taken up.

5.2.1.5 Health and Disease

Ajnauthi and Badawali has problem of malaria, diarrhea, jaundice, hepatitis and skin infections. Villagers informed that the canal water is very polluted, but due to no choice, it is being used for irrigation. They are quite aware of the effects of poor sanitation and hygiene conditions. They don't have hospital facility even for the primary medical requirement.

Villagers praised the government campaign of Pulse polio and visit of nurse and midwife for health checkup, but this is very little than their need.

Table 5.2 Fertilizer consumption (kg/ha)

Village	1995	2008
Ajnauthi	100	200
Badawali	90	200

Source Primary survey

5.2.1.6 Fertilizer and Pesticide Application

DAP and Urea is common fertilizer used by villagers for soil fertility in both the villages. Fertilizer application rate in these villages are given in Table 5.2, which reveal that the fertilizer application rate has been doubled during 1995–2008. This application rate has no scientific basis and therefore looked-up by the KVK. The farmers use organic manure. Endosulphan is the main pesticides used by villagers. They do not use any management technique such as IPM and IPNM and have no information for these programmes. However, these techniques IPM and IPNM are advocated for crop management and sustainable agriculture by Department of Agriculture and Irrigation Department. But these departments need a follow up for these villages also and provide training to *kolaba samiti* and villagers for sustainable agro-ecosystem.

5.2.1.7 Fuel

In both the villages, shrubs, bushes and the agriculture by-products are used for fuel. Cow dung and wood are mainly used. Very few household possess LPG but that is rarely used, as fuel wood is available, and easy access to refill the cylinders is quite difficult.

5.2.2 Socio-Economic Status

5.2.2.1 Habitation

About 1,000 people live in Ajnauthi and 900 in Badawali. Villages have 175 and 130 households, respectively. Brahmin, Baghel, Jat, Prajapati and Kumhar are the main residents in the village. Villagers are now having Pukka house and very few families have kutcha house. Villagers informed that before 20–25 years approximately all the houses were kutcha, but now status have been changed due to development of road network, employment in urban centers, and availability of canal water for irrigation though it is polluted.

5.2.2.2 Landholding

About 60 % area of village is under agriculture. Villagers have marginal to small land holding and about 15 % farmers have medium land holding. Only 2–3 % farmers come under the large land holding category.

5.2.2.3 Education

Both the village has about 45 % literate population. Ajnauthi has a middle school (up to class 8), for higher classes students have to go Chhata or Kosi Kalan. Badawali has a primary school only, for senior classes, students go to Chhata.

5.2.2.4 Occupation

In villages, farming is the main occupation. Few landless and marginal land-holding villagers work as agriculture worker. Also, many household has persons who work in cities as unskilled worker and educated one as office worker etc.

5.2.2.5 Migration

Due to uncertainty in water availability from the canal and saline water problem of tube well, villagers face acute problem of uncertain income. The villagers are now migrating to Delhi, Faridabad, and Agra for livelihood and work as labor. As they are not educated and trained with certain skills, they face lot of problem in settlement at the new places.

5.2.2.6 Status of Women

Women are mainly uneducated or have mere capability to signature. However, they are sending their daughters to study at village school and informed that they want them to study more. Women are facing lot of problem due to scarce drinking water facility. They used to walk for drinking water collection that consumes 2–3 h a day. Also they collect wood for fuel and spend 2–3 h daily. However, fuel wood collection is not a problem in village as they get it from agriculture and livestock by-products. Medical facility not available in village and nearby area is also a big problem.

Table 5.3 Crop production (kg/ha)

Village	Rice		Wheat		Sugarcane	
	1993	2008	1993	2008	1993	2008
Ajnauthi	2,470	3,705	1,976	2,964	6,175	11,115
Badawali	2,470	3,458	1,976	2,964	4,940	11,115

Source Primary survey

5.2.3 Amenities

School and hospitals: School education up to class eight is available at village level. For higher classes students need to go Chhata block. This leads (mainly girls) deprived of secondary education. There is no hospital in the nearby area and people go either Chhata or Kosi Kalan for medical consultation.

Electricity: Both the villages are electrified, but electricity supply is only 5–8 h a day. Insufficient electricity always disturbs irrigation through tube well.

Roads and infrastructure: Villagers showed their satisfaction for road conditions. They have *Pacca* roads up to village and also on the canal runway.

5.2.4 Agriculture and Irrigation

5.2.4.1 Agriculture

Villagers prefer cropping of wheat followed by sugarcane and rice because of market availability. Maize, mustard, barley, pigeon pea and vegetables are sown on low and least priority. Villagers sell their sugarcane to sugar mill and also make *gurh* for self use and selling in local market.

In recent years, scanty and delayed rainfall caused problems in agriculture production. Under such circumstances, farmers are able to cultivate only 10 % of their land. Out of which they claim that only 10 % will survive.

Villagers informed that this area was once very productive and variety of crops was grown but because of better and reliable income and for self use they prefer to grow wheat. These villages need institutional arrangements for crop intensification and diversification techniques which can be done by combined efforts of Department of Irrigation and Department of Agriculture for assured income and sustainable utilization of resources. Table 5.3 shows the production growth of the main crops in the villages.

The overall production has increased due to better utilization of seeds and fertilizer. Based on the secondary data (1995–2008), the seasonal cultivated areas under different crops are shown in Figs. 5.4, 5.5, 5.6, and 5.7.

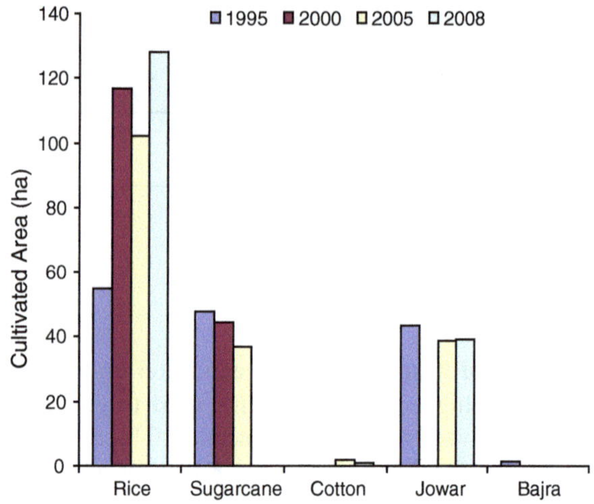

Fig. 5.4 Cropped area during Kharif season in Ajnauthi

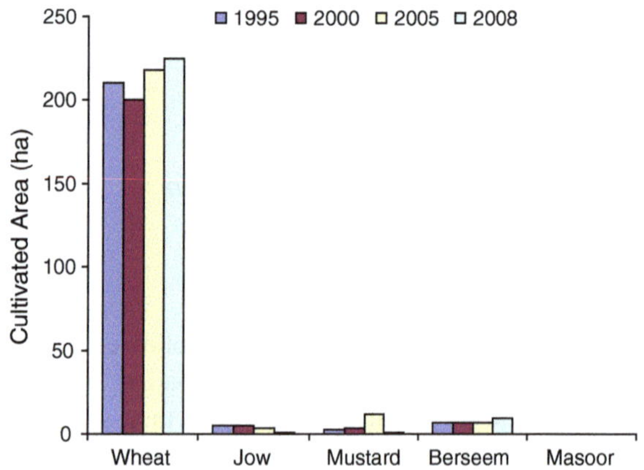

Fig. 5.5 Cropped area during Rabi season in Ajnauthi

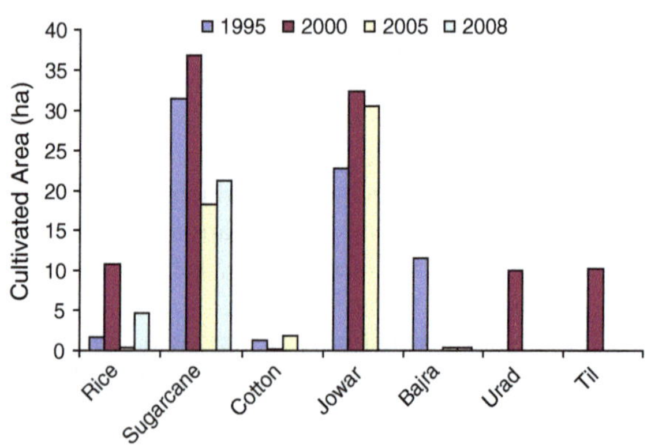

Fig. 5.6 Cropped area during Kharif season in Badawali

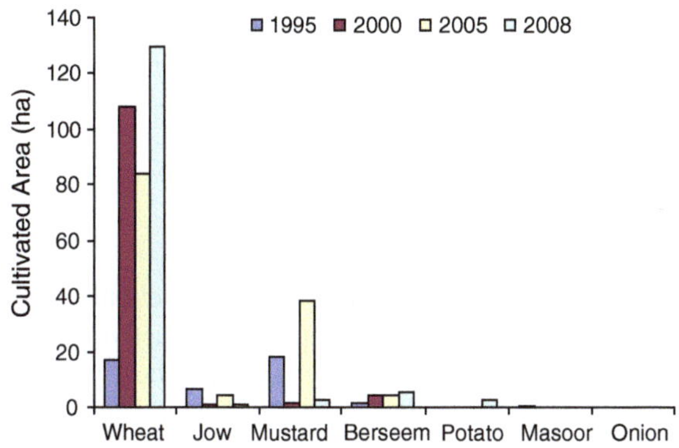

Fig. 5.7 Cropped area during Rabi season in Badawali

5.2.4.2 Irrigation

These villages come under the command area of Ajnauthi minor of Shergarh distributary (Agra Canal). Canal water supply is usually for a week in 2 months. People get turn wise water from this. *Kolaba samiti* (water user association, WUA) works for distribution of water. Villagers informed that water is supplied on the availability and as decided by the *samiti* which is contributory of Irrigation Department. Canal supplies very polluted, black and sluggish water. In Ajnauthi due to siltation slug deposition, flow is reduced and needs immediate desilting. Ajnauthi minor irrigates about 10 % of the total agricultural fields.

Villagers are in urgent need of canal water supply growth because of saline tubewell water. They highly depend on the polluted canal water, however, they are aware about the implication of such irrigation but do not have other choice. From canal they get water for 1 acre in 8 h. They have 10–12 tube wells for irrigation. They informed that 5–6 tube wells are not being used because of saline water.

5.2.4.3 Livestock

Both the villages have about 1000–1200 cow and buffalo each. Very few families are having goat and sheep (mainly the marginal landholding and landless). Very few families keep bullock also. Villagers informed that they rear the milch animal for milk production for self use and sell in the market at Chhata and Kosi Kalan. There is no cooperative federation which supports or collects milk from the village.

5.2.4.4 Farm Power and Machinery

Earlier villagers were dependent on bullock. Presently, 15–20 households have tractors. Nationalized banks in the block provide loan for tractors. Villagers informed that, this year is going to be a bad year for agriculture because of delayed rain. Villagers are worried about the payments of their loans for tractor.

5.2.5 Institutional Status

WUAs for Ajnauthi minor are not functioning. Villagers informed that these associations are merely on paper and no active participation is seen for the village development. They have never attended meeting of *samiti* and have no knowledge of its rule and regulation. There is urgent need to revive the local institutions and train them for the sustainable development. Irrigation Department and Department of Agriculture should take an initiative for better management of water and agriculture in the area. Villagers showed interest for management practice and training for better crop intensification and diversification.

A study commissioned by the Rajiv Gandhi National Drinking Water Mission (RGNDWM) to ascertain the impact of government investments on sustainability of drinking water and sanitation services in six villages of Chhata block of Mathura district has come up with the following findings:

- Piped water supply schemes, planned without consideration to groundwater conditions and socio-economic conditions, are facing threats due to brackishness/salinity in ground water and poor utilization. Where possible, people have fallen back on the use of open (unprotected) dug wells.
- There are no recognized user groups such as Village Water and Sanitation Committees or any sub-groups to the panchayats to oversee the workings of water supply and sanitation facilities.

These villages should reduce emphasis on cereals and enhance marginal emphasis on pulses, oil seeds, sugarcane and high emphasis on livestock. Institutions should work for diversification and intensification for sustainable ago ecosystem.

5.2.5.1 Capacity Building

Local level institutions may play major role for development and training at villages. *Samiti* at village may be trained for better cropping and irrigation. They may be trained for better management of fertilizer and pesticides through integrated nutrient and pest management. Villagers may be trained for pasture, social forestry and farm forestry for assured income from livestock.

Table 5.4 Summarized status and possible recommendation of the issues

S. No.	Issue	Status	Concluding remarks/Recommendation
1	Education	Inadequate amenities schools: no secondary school	Secondary school as well as encouragement for girl education is needed. Vocational training required for off-season income
2	Drinking water	Drinking water scarcity and salinity problem	Secure drinking water supply is required, as most of the dug/tube well have the salinity problem
			Piped water supply from nearby location is needed. Due to enhanced desalination technology, salinity can be removed at least to ensure drinking water supply. More generally, desalination may also refer to the removal of salts and minerals as in soil desalination
			Salt water is desalinated in order to produce fresh water that is suitable for human consumption or irrigation. One potential by-product of desalination is table salt. Most of the modern interest in desalination is focused on developing cost-effective ways of providing fresh water for human use. Along with recycled wastewater, this is one of the few non-rainfall-dependent water sources
		Ground water (GW) level is lowering	GW utilization at sustainable basis should take place in that area
			For increasing GW recharge a simple technique that could be incorporated by the villagers themselves is diverting rain water into a recharging wells or injecting wells through either an underground pipe system or through an open ditch. This can be constructed by help of local institutions and public participation. MNREGA may help in creating percolation tank in the area at suitable site
3	Sanitation and hygiene	Open defecation	Community toilet may be proposed
		Waste disposal problem	Disposal of domestic wastewater is also a big problem in these villages. Soak pits shall be usable for these areas, which are simple and cheap drainage structures facilitating the disposal and filtration of wastewater. This can be constructed by help of local institutions through public participation
			Disposal of waste should not be inside the habitation as it may generate communicable diseases
			Disposal sites should be low land area and barricaded so that any seepage/leachate from this will not meet with natural streams. Also it should not pollute the ground water
			As these villages have livestock, so biogas plant will be beneficial for fuel as well as reducing problem of waste
			Based on the landuse and topography, suitable solid waste disposal site can be developed along with recycling plant

(continued)

Table 5.4 (continued)

S. No.	Issue	Status	Concluding remarks/Recommendation
4	Electricity	Frequent power cuts (5–8 h supply). Power cut for continuous 2–3 days is common	Daily supply for higher duration is required to fulfill the irrigation and home use water demand Alternate energy sources like solar power should be promoted by the government. It will not improve the livelihood but can help in improvement in the irrigation efficiency
5	Land resources	Soil is losing productivity; villagers claimed that it is due to saline water from tube well and polluted water of canal	Department of agriculture and Irrigation Department may intervene and suggest/work fruitfully
6	Forest	The forest area of village has been taken by the forest department	Ajnauthi and Badawali are deprived of forests. Forest department has also not worked in/or the nearby area. These villages should have forest and pasture for livestock grazing and other by products such as tree lopping etc. for livestock feed. Approach should be to work for social forestry/farm forestry so that besides development of green, certain security for livestock feed and fuel wood is also developed
7	Water resources	Only 1–2 small ponds exist in village	As the area is flat and deep aquifer is saline, it needs storage of rain water for recharging the upper aquifer Low land areas should be identified and developed for water harvesting
8	Agriculture	Low crop yields Increasing cultivation cost Reduced crop diversity	Soil conservation measure and usage of organic manure should be increased Crops with low water demand should be promoted. This will reduce the cost of cultivation and over exploitation of ground water Crop diversity should be enhanced for secure economy of village as well as better crop-water management
9	Irrigation	Tube well Salinity problem Canal: dirty blackish wastewater	Very few tubewell provide sweet water. Canal water is very dirty and black in color, and villagers are left with no choice. Irrigation department may take action on water quality of canal, as this water is deteriorating the soil and groundwater and ultimately the health hazards. As the canal water is disposal of sewage treatment plants mainly, this may have toxic elements, as it comprises domestic as well as industrial effluent
	Irrigation water quality	Poor irrigation water quality. Adequate disinfection of the effluent required Canal water (effluent) must meet normal irrigation water requirements for parameters such as salt content, sodium adsorption ratio, trace elements, etc Long-term effects of sewage irrigation on underlying groundwater need to be ascertained	Sewage irrigation has main concern for its potential for infectious diseases in farm workers as well as in people who consume crops irrigated with effluent, especially when those crops are eaten raw or brought raw into the kitchen Since most of the water applied for irrigation in dry months evaporates, the concentrations of non-biodegradable chemicals in the drainage or deep-percolation water going down to the groundwater can be much higher than in the effluent itself Thus, groundwater below sewage-irrigated areas eventually may become unfit for drinking, which raises questions of liability

(continued)

Table 5.4 (continued)

S. No.	Issue	Status	Concluding remarks/Recommendation
10	Livestock	Fodder management	Livestock are major part of rural economy. Villagers grow fodder crop and provide by-products of agriculture to their livestock. Need of protected pasture is urgent in these villages Plantation of fodder tree needs to be prioritized Protected pastures and farm forestry may be developed as these villages have tremendous scope for livestock economy
11	Non farm sector	Not scaled up	Non-farm activities may include small scale or in house work for revenue generation. Basket making, embroidery etc. can be easily done by women. However, this can only be done once they achieve security for drinking water; they spend lot of time on collection of drinking water and fuel wood
12	Economic security	No stabilized income Risk during aberrant weather Irregular and dirty water supply from canal	Crop diversity, pasture, farm forestry and non- farm activities should be implemented for stabilize income. As the area is at high risk of irrigation water, damage of crops and soil may lower the production
13	Institutions	Water user association (*Kolaba samiti*)	Samiti has limited scope for water distribution from canal Samiti may be trained to train villagers for better livestock rearing, farming and non-farm activities for revenue generation Samiti may contribute through proper training for villages regarding integrated nutrient and pest management and suitable cropping pattern The power of Samiti is limited; these may be used for broader perspective for sustainable agro ecosystem
14	Landscape	Landscape transformation	Awareness about the changing effects of humans on landscapes with shrinking forests, mismanagement of cropping and irrigation, degraded agriculture fields may serve as a useful tool in formulating regionally appropriate policies for conserving agro ecosystem

5.3 Summary and Conclusions

Villagers in both the villages have desire for development but they are not getting support of the local administration. They also claimed that they are deprived of developments as the administration favours only the bigger *Panchayat* villages. Based on the information collected from these villages, an overall view for this region i.e. Middle Alluvial Pains can be summarized along with the issues and possible recommendations (Table 5.4).

Local public are already involved in various programmes under MNREGA of Government of India. In a similar fashion, if government takes initiatives by convincing the principle components of IWRM i.e. Sustainability, Environment, Economy and Equity, public will be equally participate as they are keen to solve their problem.

References

CGWB (2010) Groundwater status of India. Central Ground Water Board, Faridabad, India

Rai RK, Upadhyay A, Ojha CSP, Singh VP (2011) The Yamuna river basin: water resources and environment. Water Science and Technology, vol 66. Springer, The Netherlands

Chapter 6
Case Study: Banganga Catchment

Abstract The Banganga catchment comes under semi-arid climate and has scanty annual rainfall. Excessive groundwater exploitation has further aggravated the water problem. The catchment has rich mining resources and therefore, extensive mining work without their cumulative impact study has largely affected the catchments' water resources, especially the ground water. In the area, NGO has done extensive work for water resources development with public partnership using the indigenous practices, and succeeded remarkably. The lesson learned from this catchment can be replicated to other parts of the basin having similar hydro-geological and climatic features.

The Banganga River originates in the Aravali hills, near Arnasar and Bairath in Jaipur District, and flows towards the south up to the village of Ghat, then east through partly hilly and partly plain terrain. The total length of the river is 240 km. The catchment of Banganga River is located in the northeastern part of Rajasthan, between latitudes 26°40′ and 27°37′ and longitudes 75°49′ and 77°39′. It lies between the Gambhir and Banas river catchment of Chambal sub-basins to its south-southwest, Ruparail and Sabi to its north, and the Shekhawati catchment to its west. Its eastern edge borders the Yamuna River basin in Uttar Pradesh. Banganga River Basin extends over parts of Alwar, Jaipur, Dausa, Sawai Madhopur and Bharatpur Districts. The drainage area of the Banganga catchment is 8,878 sq km. Orographically, the western part of the Basin is marked by hilly terrain belonging to the Aravali chain, with fairly flat valleys along the Banganga River and its tributaries. East of the Todabhim–Mandawar chain of hills lies an extensive alluvial plain which gently slopes eastwards, towards the Yamuna River in Uttar Pradesh. The northeastern part of the area is also rather flat, interspersed with moderately elevated hills (Jain et al. 2007; Rai et al. 2011).

The mean annual rainfall over Banganga Basin was computed as 596 mm, of which about 95 % fall during the four Monsoon months (June–September).

A. Upadhyay and R. K. Rai, *Water Management and Public Participation*, SpringerBriefs in Earth Sciences, DOI: 10.1007/978-94-007-5709-7_6,

Catchment is vulnerable to the flooding under extensive rainfall. In the area, increasing groundwater use has led to depleted aquifers. In such case, the rainwater harvesting is the only feasible technique to augment the surface and groundwater storages. Socio-economic as well as ecological gains of rain water harvesting has been highlighted in many areas. An extensive work has been done regarding the water harvesting and natural resource management in Thanagazi block of Alwar district. Three villages Bhanwata, Samra and Ajabgarh were studied for assessing the status of water conservation and its impact, and public participation. Villagers were discussed and secondary data from *Tehsil*[1] office was also collected.

Bhanwata and Samra village has done remarkable job for water harvesting and conservation by constructing *anicut, johad, medbandi* and plantation, which helped in restoring the groundwater for round the year use. Farmers improved the crop production under scanty rainfall and drought condition. Bhanwata village received award for this exemplary work for restoration of Arvari River. Ajabgarh being a princely area benefited by the big concrete dam named as Jai Sagar constructed in by ruler of Alwar, *Maharaja Jai Singh* about 100 years back. Now it is governed and maintained by the State Irrigation Department. Maharaja *Jaswant Singh* the grandson of founder *Maharaj Ajab Singh* built a wall across the valley separating Ajabgarh from Bhangarh (a nearby fort). He did an important role in fortifying the area. Part of old Ajabgarh seems curfew area in the walled portion, because of some *Tantric*[2] practice which killed many villagers and forced to evacuate.

6.1 Geo-Physical Characteristics

6.1.1 Climate and Rainfall

Climate of the area comes under the semi-arid with hot in summer and very cold during the winter. The mean annual rainfall over Banganga basin is 596 mm, of which about 95 % fall during the four Monsoon months June–September (Water Resource Department, Rajasthan).

6.1.2 Topography and Soil

Bhanwata and Samra village fall under watershed boundary of Arvari River. Dominant landscape features are flat topped hill and inter ridge valleys with plains stretching southward. The average height of the ridges ranges from 640 to 690 m

[1] Taluka or tehsil is part of a larger District within a State. In some cases, the tehsils of a given District are organized into smaller groups of tehsils called Sub-Division.

[2] As informed by villagers (however, it seems may be due to some disease people were killed).

in the central part of the area. The southern and eastern areas have low to moderately high hills (500–600 m) with low gradients. The geological sequence generally consists of limestone, quartzite, phyllites, alluvial sediments and wind-blown sand, silt and clay. Topography in Bhanwata is undulating with hills and hillocks and few plain patches which have been developed for agriculture. Soil depth is low. In Samra also terrain is undulating. Ajabgarh has better land feature for agriculture. Soil within the rain water storage systems belongs to sandy loams to light medium clays with shallow depth.

6.1.3 Biodiversity

Common plant species in the selected villages were *Cassia fistula, Anogeissus pendula, Acacia nilotica, Butea monosperma, Zizyphus nummularia, Z.mauriiana, Dalbergia sissoo, Azadirachta indica, Nerium indicum, Euphorbia cadicifolia, Rhus misurensis, Hardwickia binata, Prosopis cineraris, P. juliflora, Bauhinia racemosa, Wrightia linctoria, Salvadora oleoides and Mangifera indica.* Fauna species in the nearby forest area commonly found are, wolf, deer, rabbit, Neel gai, monkey and ape.

6.1.4 Common Crops and Cropping Pattern

The agricultural land use has three principal cropping seasons, *Kharif*, *Rabi* and *Zaid*. *Kharif* crops are predominantly maize, pearl millet, sorghum and pulses. *Rabi* crops are wheat, mustard, barley and black lentils. *Rabi* is irrigated entirely with groundwater. The *Zaid* or summer crop is mainly vegetable; it is only planted if there is sufficient groundwater.

6.2 Inferences from Primary Survey

Through primary surveys, various information about the past and current situation of water related issues have been collected. In the area various water harvesting structures have been constructed through public participation, which considerably manage the water problem in the region (Figs. 6.1, 6.2, 6.3, and 6.4, 6.5, 6.6). However, water harvesting structures constructed in the area are not so effective because of lack of technical understanding. It was also appeared that the new constructed structures have caused unavailability of water to meet the installed capacity of the existing reservoirs or structures. Therefore, while developing any new structures, cumulative impact on the other water resources development project should be evaluated. Apart from the primary survey, focus group

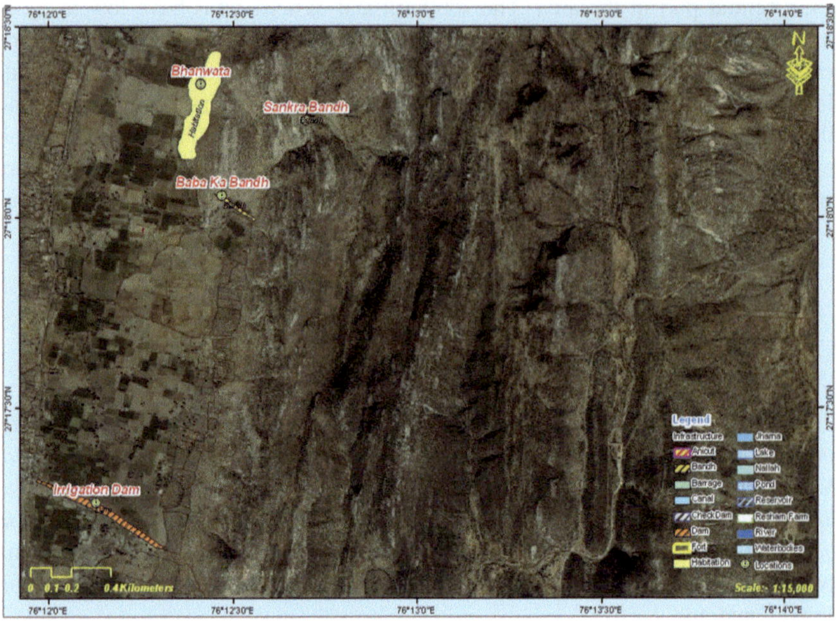

Fig. 6.1 Major WHS and habitation in Bhanwata

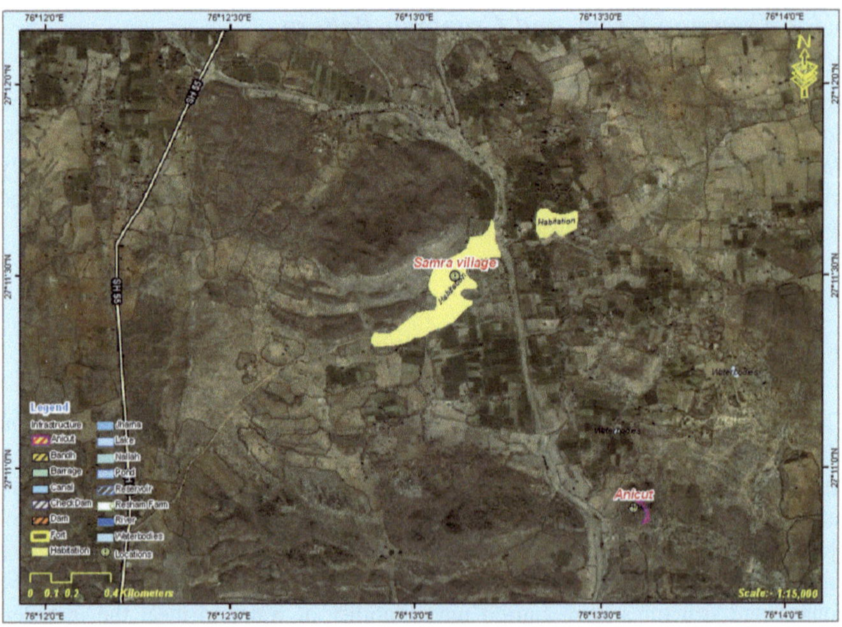

Fig. 6.2 Major WHS and habitation in Samra

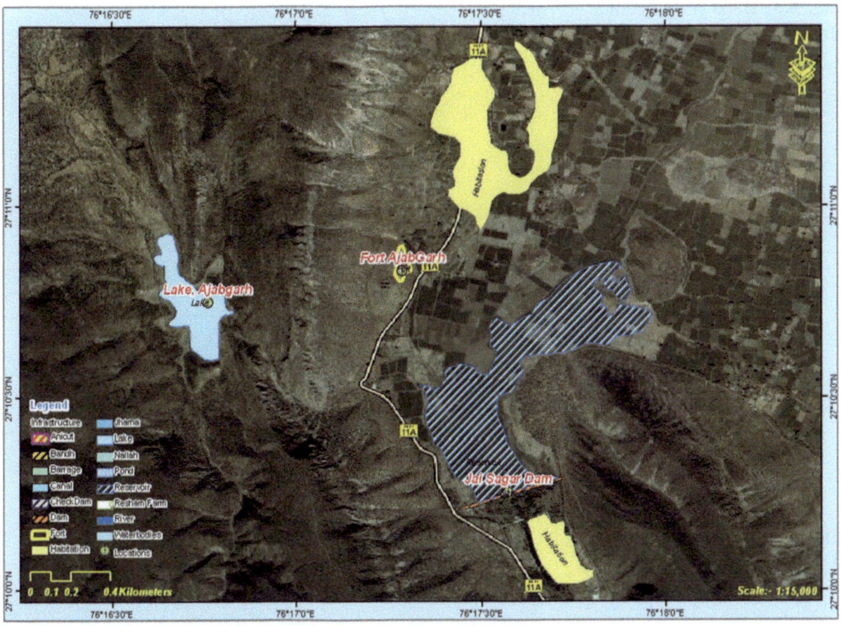

Fig. 6.3 Major WHS and habitation in Ajabgarh

Fig. 6.4 Sankara bandh, Bhanwata, Alwar

discussions have been made to confirm the real status of water resources issues and their impact on the socio-economic (Figs. 6.7, 6.8, 6.9, 6.10, 6.11).

Initiatives of non-government organization (NGO) in the area for water harvesting are remarkable. The initiatives were taken after a severe drought in Alwar district in Rajasthan during 1985–1986. In 1985, volunteers from the Tarun Bharat Sangh (TBS), a voluntary organization led by Rajendra Singh, came to Alwar. The Alwar District had once thrived, but logging, mining, and other industrial activities

Fig. 6.5 Baba ji ka bandh, Bhanwata, Alwar

Fig. 6.6 Jai Sagar dam, Ajabgarh, Alwar

resulted in land degradation that intensified flooding and droughts. In the area, the traditional water management system like *Johads* (water storage tank built for surface and ground water augmentation) was abandoned. TBS revived these traditional systems and currently popular as Tanks in most parts of the country. The NGO especially the TBS has also contributed to develop water awareness to the people, and efforts have been made in such a way that local public have equally participated in the water resources development for their socio-economic growth. Currently, it can be seen that the water harvesting is inherent in Bhanwata, Samra and Ajabgarh as village culture. Village has traditional harvesting system. Bhanwata has about 5 community based water harvesting structures. Samra has 20 *Johads* (i.e. water storage structure) and 1 *anicut* (weir constructed across the river). In totality, Samra has 21 community-based and 19 private owned water harvesting and soil conservation structures. Ajabgarh has one pond and 27

Fig. 6.7 Women working at Bhikampura, Alwar under Scheme

Fig. 6.8 Imminent conservationist of Bhanwata village, Alwar

Fig. 6.9 Meeting with legends of Bhanwata, Alwar (Dhanna baba contributed in water safeguarding)

Fig. 6.10 Meeting with legends of Bhanwata, Alwar (Arjun baba played active role in consolidation of water crusade)

Fig. 6.11 Open well in Agar village, Alwar (well was dry at the time of visit)

Johadis. Water harvesting and conservation measures increased the agriculture land use and crop production. Nomenclature of these traditional structrures is defined as follows:

- *Johad*: *Johads* are earthen dams of semi circle shape (Fig. 6.12). These have been constructed at the foothills of slopes for collecting water from small hilly catchment area. The main purpose of *Johads* is for livestock drinking water with some contribution to recharge.
- *Johadis*: *These are* small *Johads*.
- *Talab*: *Talab* is a pond like structure that has high raised edges on 3 sides made of earth (similar to *Johad*). Monsoon water is collected in the *Talabs* for village use and livestock drinking with some small contribution to recharge.
- *Talais*: *Talais* are *are* smaller *Talabs*.

Fig. 6.12 Typical view of Johad in Rajasthan

- *Medbandi*: *Medbandis* are constructed in agricultural fields. The lower sides of the field are raised slightly to retain runoff to increase soil moisture content, and are made from earth. The purpose of a *Medbandi* is to retain agricultural runoff.

In Bhanwata village, river Arwari originates behind Sariska hill area. Arwari watershed area is 503 sq km. and river length is about 45 km. it culminates in Santhal Sagar dam near Neemla village. Another *nallah* coming from Hamirpur and Thali village culminates here. As such with this name a very small *nallah* exist in Bhanwata, in other villages the *nallah* have different name or no name even. Arwari is more known to Bhanwata villagers.

Bhanwata village was facing huge water crisis, TBS has motivated and helped villagers for participation in community work. They helped in construction of rainwater harvesting structures and *medbandi*. Since then Bhanwata has learned to stand on its own and they have developed capacity to survive under deficit year (i.e. when actual rainfall is less than the normal). Villagers informed that if they get good rain even in one year it helps up to three years through these structures by groundwater recharge. Villagers of Samra informed that before 30 years the small drains and *nalla* in the village used to flow[3] for 7–8 months in a year; which has been reduced to 4–5 months only. Presently, stream flow has become limited for 1–2 months during the monsoon because of extensive mining work. Samra is downstream to Bhanwata, this may be a reason that water harvesting at upstream has checked the downstream flow. However, studies indicate that these structures have actually helped the ground water augmentation. In Samra also villagers informed that good rainfall in 2008 monsoon which helped in recharging the water level helps 2–3 years. This is an exemplary model which people of these villages have done. Villagers also informed that if *Samar Sarovar* site near Jhiri (earthen

[3] This indicates the gradual change in climatic patterns and its impact on stream flows.

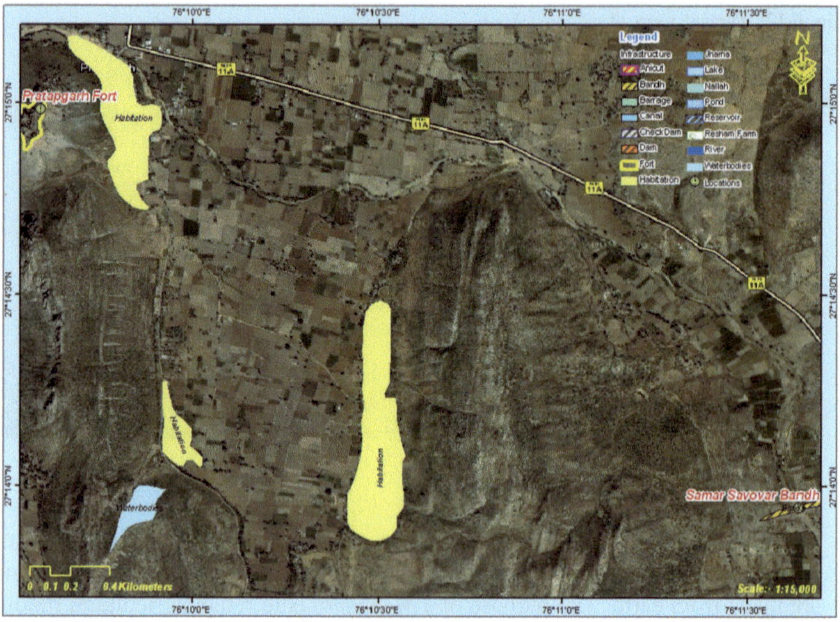

Fig. 6.13 Habitation and location of Samar Sarovar site near Jhiri, Alwar

embankment was constructed but breached due to flooding) is reconstructed for water conservation, it will benefit to the area of 6 villages. The location of the Samar Sarovar site is shown in Fig. 6.13. However, construction of this dam requires financial help which comes under jurisdiction of Irrigation Department. For this purpose, villagers are approaching administrative and political circle to get listened. Near Jhiri, drinking water supply pipe has been laid but the tube well to provide water is not functional since the beginning of the project, and this area is categorized as piped water supply by PHED.

Water harvesting structures available in the study area:

- Weir: Weir is a low head overflow masonry structure built in small tributaries.
- Small dam: An earthen embankment with a spillway built in major tributaries.
- Anicut: Anicuts are built across the main n*allah.* They are generally *pacca* structure (made of cement, stone and concrete). Their main purpose is to recharge the aquifer.
- *Bandh*: *Bandhs* mainly sit across tributaries of the main river. They generally consist of a concrete core, but the outer edges are of earth, and some are entirely made of earth. The main purpose of a *Bandh* is recharge.

6.2.1 Role of Local Institution

A small institution, *Gram Sabha* was made at village level which comprise of 16 people in Bhanwata and 25 in Samra for the maintenance and repairing of structures and for the follow up of the rules for cropping pattern and natural resource management in light of the Arwari Sansad. [4] *Gram Sabha* also comprises of *Gram kosh* which acts as seed bank; however, at present, it is not functional. In these villages involvement of women is poor at institutional level; however, they actively participated in the development of water harvesting structures. These rules have stopped cutting of trees and villagers collect dry twigs fallen on ground only. TBS also did afforestation work and developed protected area, which helped to rejuvenate capacity of the forest land in the degraded hilly and hillock patch of Bhanwata.

6.2.2 Environmental Status

6.2.2.1 Water Table

Bhanwata village has shown increase in water table in dug well after these interventions. There are about 20 wells in the village. Samra has about 200 well. Villagers have done lowering of depth in many wells for enhancing water availability. At the time of field visit also, deepening of 4–5 wells was going on. They have 160 diesel engine pumping sets for groundwater abstraction. In Samra, it was informed that before 30 years back, 25 *rahat* along with wells were exist. Gradually decline in the water level started from last 20 years mostly because of the mining work. At present, only 25 wells are working which are used for drinking water and irrigation. Presently, average water table in the area was 22.0–24.0 m. Ground water declination is common in Ajabgarh village also, however, 40 open well exists.

Table 6.1 shows the impact of intervention of water harvesting structures on groundwater development. Water level in the dug well has shown impact of the water harvesting structures through good rainfall, which is clear in Bhanwata and Ajabgarh. It is clear from the Table 6.1 that the groundwater levels in the dug wells have been improved during the period 1990–1995. However, during the year 2001–2010, groundwater has been further depleted due to agricultural growth as well as scanty rainfall (Fig. 6.14).

In case of Samra village, impact of rainwater harvesting structure on groundwater is not apparent in the Table 6.1; however, agricultural growth has been

[4] Arwari Sansad is informal institution which belongs to villagers of Arwari catchment. It has framed rules and regulation for better practice of integrated water resource management. This is brain work of Tarun Bharat Sangh (TBS), an institution headed by Magsaysay award winner Mr. Rajendra Singh.

Table 6.1 Water level in open well in the selected villages (mbgl)

Village	Year 1990		Year 1995		Year 2001		Year 2010
	Pre monsoon	Post monsoon	Pre monsoon	Post monsoon	Pre monsoon	Post monsoon	Pre monsoon
Bhanwata	26	20	18	13	20	21	23
Samra	26	24	32	32	33	32	28
Ajabgarh	25	18	24	17	26	21	23

Source Primary survey (as informed by villagers)

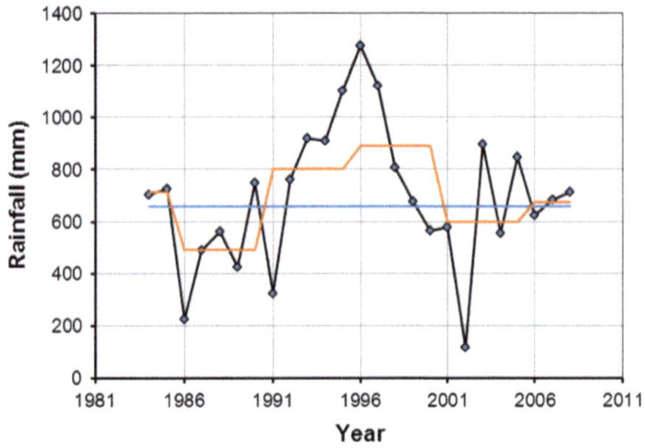

Fig. 6.14 Rainfall in Thanagazi block of Alwar (*blue line*: mean rainfall, *orange line*: 5 years consecutive mean rainfall) (*Source* Irrigation Department, Rajasthan)

improved. This might be due to inappropriate site selection for harvesting work, which was apparent during the site visits. In Ajabgarh, water table fluctuations are low and status is steady. Further developments of rainwater harvesting structures were not done; only old ones have been improved. It is near Jai Sagar, which directly affects the village.

6.2.2.2 Sanitation and Hygiene

The open defecation system is generally applied by the villagers. In Samra village an NGO tried to work for toilet construction but it has not been taken over by villagers and these are abandoned because of water scarcity. There is no facility as such in the area of the catchment. However, cleanliness concept is well adopted. We do not find wastewater or garbage thrown anywhere messily in the selected villages.

6.2.2.3 Solid waste Disposal

Villagers perform open dumping and area is barricaded with bushes. This avoids it spreading through air and water. By-products of agriculture are used for animal feed and which is not usable for fuel and dumping for degrading to use as organic matter for agriculture.

6.2.2.4 Disease

Even being in the water scarce area, the villagers perform cleanliness that has helped in avoiding general diseases which occur due to wastewater and contamination. Swasthya Kendra is available in nearby area and auxiliary nurse & midwife (ANM) also visit village. Pulse polio scheme has also been successfully adopted which received appreciation.

6.2.2.5 Fuel

Women were seen collecting dry twig in the hilly area near Sankara bandh. Usually women take their goats and sheep in near by hilly area which allows fodder to livestock and fuel wood. Prosopis juliflora has been planted to get fuel wood in the village.

6.2.3 Socio-Economic Status

6.2.3.1 Habitation and Landholding

Bhanwata has population of 419 with 50 household. Operational land holding in Bhanwata is marginal to small. However, Samra village is more populated and bigger than the Bhanwata, having population of 2,195. Village has group of 15 *dhanis*[5] such as *Tigria ki dhani, Gyaran ki dhani, Deo ka devra, Nizara Nichli, Van ki dhani* etc. with 370 household. Operational landholding in Samra is marginal to medium. About 20 household come under landless category. Ajabgarh village is very old village and was very influential in the time of *Kachhwas kings.* Ajabgarh has population of 1,420 with 200 households. *Gurjar, Rajput, bania, meena* and *balai* are the main caste category live in these villages.

6.2.3.2 Education

Bhanwata village has primary school. TBS volunteers also helped here in educating people. Samra and Ajabgarh also have government school. Students can study up to middle class in the village itself. *Bodh Sanstha* a non-government

[5] *Dhanis* are habitation, in nearby agricultural field and water resources.

organization has also worked for literacy in the villages. About 30 % population in Bhanwata and Samra is literate.

6.2.3.3 Occupation

Bhanwata and Samra were earlier household of shepherd families generally involved in cattle rearing with very little agriculture. Because of the water conservation and soil conservation measure since long time they developed agricultural land and started farming. Now farming is first occupation followed by agricultural labourer and cattle rearing. However, Ajabgarh was mainly agricultural based village. Main occupations of the villagers are mainly farmers and agricultural labours. Some villagers have got employment in nearby marble mines in Pratapgarh and Jhiri area.

6.2.3.4 Migration

Water scarcity for agriculture and fodder availability does not influence the migration in these villages. Villagers have developed coping mechanism through water harvesting which help them in surviving under rainfall deficit year. Household with marginal land holding and landless migrate to Alwar and Delhi for masonry work.

6.2.3.5 Status of Women

Women were seen front runners in these villages. They do household work, drinking water collection, fuel wood collection and roaming for grazing of animals. They actively participate in agriculture. Women are mostly illiterate; however, the new generation is getting education. They have respect for education and consider illiteracy as their ill fate. Women informed that before these development work for water and plantation, situation was very difficult even for survival. They had to cover 1–2 km for fuel wood collection.

6.2.4 Agriculture and Irrigation

Bhanwata has total geographical area of 221 ha, out of which 133 ha (i.e. 60 %) land is under forests. In the village, net cultivated land is approximately 90 ha (i.e. 40.7 %), in which only 36.6 % (i.e. 33 ha) is irrigated through the dug wells as only source. Samra on the other hand is a large village having topographic

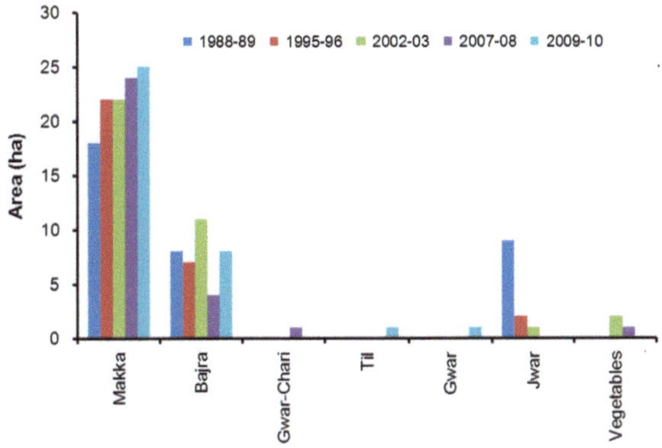

Fig. 6.15 Cropped area during Kharif in Bhanwata

extent of 2057 ha, in which forest and total uncultivable land comprised an area of 618 and 1,048 ha (50.9 %), respectively. In the Samra village, only 23.9 % (i.e. 492 ha) area is suitable for cultivation in which net sown area is 58.1 % of cultivable land (i.e. 286 ha). The net irrigated land in the village is less than half of the net sown area (i.e. 133 ha). For irrigation dug wells are the major source.

The net cultivated land in the Ajabgarh village is 161.13 ha which is 31.22 % of total geographic area (i.e. 516.0 ha). Major part of the area is under forests and hills. Villagers informed that crop production has increased because of availability of water in Bhanwata. However, over all it has increased in Samra and Ajabgarh also which is due to better farm practices. In all the three villages, two principal crop season is distinguished (i.e. *kharif* and *rabi*). Cropped area in these villages for *rabi* and *kharif* are shown from Figs. 6.15, 6.16, 6.17, 6.18, 6.19 6.20. *Zaid* crop is also taken in Ajabgarh.

Main kharif crops in Bhanwata are *Makka*, *bajra* and *jwar*. In Rabi wheat, *jau*, *chana* and *sarson* are main crops. Kharif crops are mostly un-irrigated except vegetables. Cropping area under *makka* for kharif and wheat for rabi has increased. However, cropped area under *jwar* has been reduced till the crop calendar year of 2002–2003 and after that it has been dropped (Fig. 6.6). Crop diversity is less in Bhanwata as compared to Samra and Ajabgarh. M*akka* and *bajra* are the main *kharif* crops of Samra, in which Tobacco is the only irrigated crop. In *rabi* all crops are irrigated in these vilages.

In Ajabgarh village, farmers are aware of the adoption of cropping pattern according to the availability of water. During the kharif season of crop calendar year 2009–2010, crops adopted for cultivation are *Bajra*, *tilhan* and *makka*, whereas, during the years 2007–2008 and 2008–2009, crop order was *makka* (maize), *bajra* (barley) and *tilhan* (oil seeds). In rabi, wheat has been slightly reduced while cropping area under *sarson* (mustard) and *dhania* has increased.

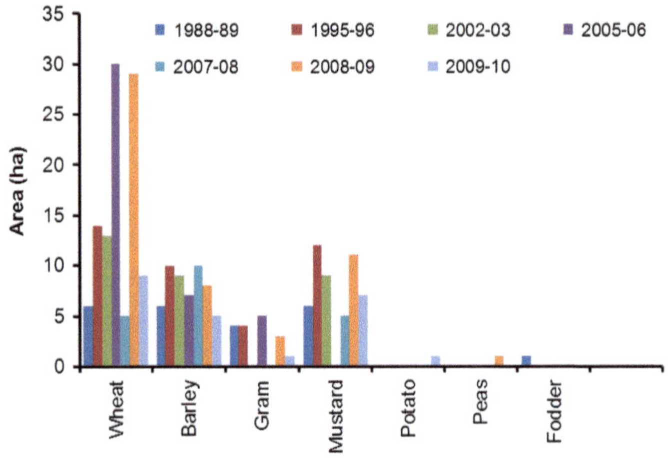

Fig. 6.16 Cropped area during Rabi in Bhanwata

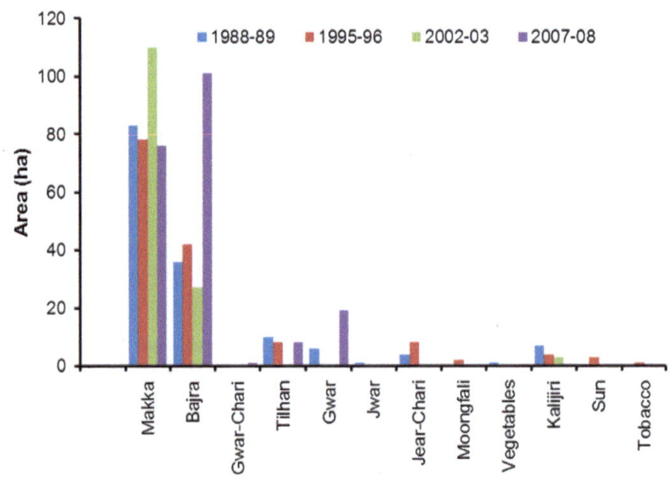

Fig. 6.17 Cropped area during Kharif in Samra

6.2.4.1 Farm Power and Machinery

Tractors and bullocks are the main power source of farming used for ploughing. In Bhanwata, farmers are either small or marginal. There is only one tractor available with Bhanwata village, however, Samra village has 7 tractors.

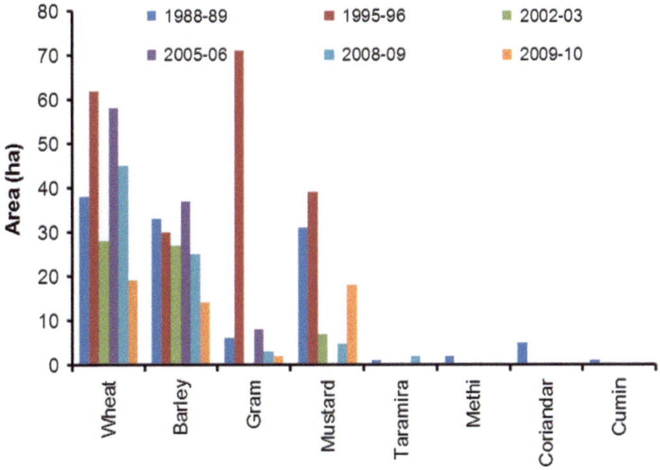

Fig. 6.18 Cropped area during Rabi in Samra

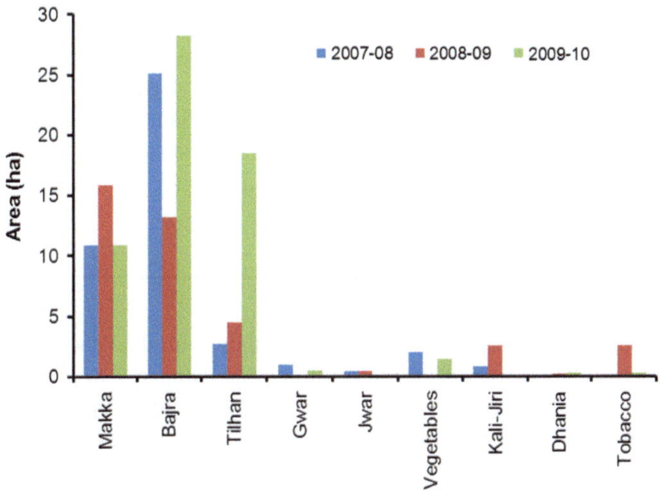

Fig. 6.19 Cropped area during Kharif in Ajabgarh

6.2.4.2 Irrigation

In Bhanwata and Samra, dug wells are the major source of irrigation. There are 6 dug wells in Ajabgarh which are used for irrigation. Constraints to agriculture in these areas include low, seasonal, high intensity, erratic and localized rainfall, and high evaporation, shallow depth of soils (which tend to be highly stony and have high gravel content) and gentle to moderate slopes in the vicinity of hills. There is complete dependency over groundwater for irrigation, particularly in the *Rabi* and *Zaid* crops.

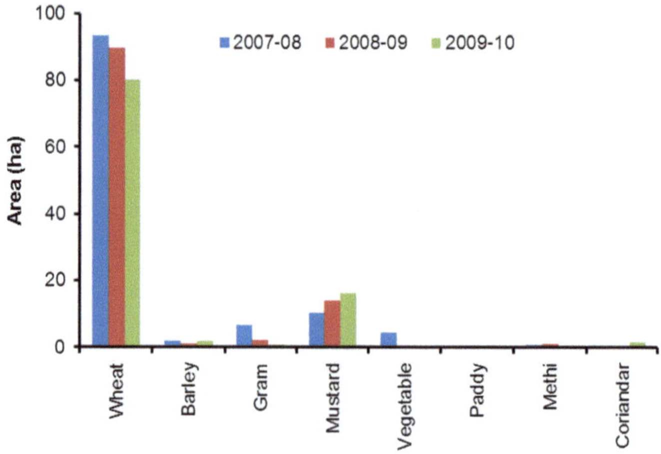

Fig. 6.20 Cropped area during Rabi in Ajabgarh

6.2.4.3 Crop Production

Villagers informed that crop production per hectare has increased because of water harvesting, utilization of high yielding variety and fertilizers. Bhanwata and Ajabgarh have better production as compared to Samra. Among these two, Ajabgarh has better crop yield due to crop-water management and better soil condition. Village-wise crop yield is given in Table 6.2.

6.2.4.4 Fertilizer and Pesticide Application

Di Ammonium Phosphate (DAP) and Urea is mainly used as chemical fertilizer. Its consumption has increased in these villages as compared to years before the development of rainwater harvesting structures. Pesticide consumption has not been changed. Villagers informed that fertilizers are available in nearby depot,

Table 6.2 Crop production in the selected three villages (kg/ha)

Crop	Year 1998			Year 2005			Year 2008		
	Bhanwata	Samra	Ajabgarh	Bhanwata	Samra	Ajabgarh	Bhanwata	Samra	Ajabgarh
Wheat	2,900	2,800	3,000	2,900	2,900	3,200	3,100	3,100	3,200
Bajra	1,150	1,180	1,200	1,150	1,100	1,200	1,200	1,160	1,280
Jau	2,200	2,100	2,400	2,300	2,300	2,400	2,400	2,400	2,500
Makka	1,200	1,250	1,400	1,200	1,250	1,500	1,280	1,200	1,500
Chana	950	800	900	900	850	900	1,000	850	950
Sarson	800	900	900	800	850	1,000	900	1,000	1,100

Source Primary survey (as informed by villagers)

Table 6.3 Livestock population in selected villages

Livestock	Year 1980	Year 2010
Bhanwata		
Cow	40	30
Buffalo	50	40
Bullock	50	20
Goat and Sheep	2,000	800
Samra		
Cow	150	125
Buffalo	200	160
Bullock	100	50
Goat and Sheep	5,000	3,000
Ajabgarh		
Cow	120	100
Buffalo	150	130
Bullock	150	40
Goat and Sheep	5,000	3,000

so they can easily get. The cost of rearing animal is high that lead to decrease in the use of cow dung as manure.

6.2.4.5 Livestock

In all the villages, livestock population has been considerably reduced (Table 6.3) due to lack of pasture land and no entry in protected forests by Forest Department, etc. Cow, buffalo, goat and sheep are the main livestock in all the villages, however, villagers prefer for goat and sheep because of non-requirement of stall feeding. Villagers keep buffalo and cow for milk production for their household consumption. In Samra, few families sell milk in the local market.

6.2.5 Amenities

School and Hospital: School is available in village and hospital facility is available within 2 km. For secondary education, students need to go Pratapgarh and Thanagazi.

Road and Transport: All these villages are well connected to *Pacca* road. Private and government buses are available to commute in nearby area. For other cities also transport facility is available via Thanagazi and Pratapgarh.

Electricity: Electricity supply has been provided to these villages, however most of the time it is not available.

Fig. 6.21 Tarun Bharat Sangh (TBS) Ashram at Bhikampura-Kishori, Alwar

6.2.6 Institution

The villagers in the Banaganga basin were facing the problem of water scarcity because of scanty rainfall. Monsoon failure is regular phenomenon in the catchment. Under drought, villagers used to migrate in lower areas for livestock feeding. Due to excessive groundwater pumping and watershed degradation because of deforestaion and extensive mining, the entire area was under problematic condition. Villagers in Alwar for more than a decade worked with a voluntary group named TBS (Fig. 6.21) to build more than 4,000 small and some large structures in villages to collect rainwater to augment groundwater storage through recharge. TBS worked here for constructing and reviving old structures by village contribution in kind or labour. Giving resource property rights to communities, so that it is managed as common property resource, has been encouraged as an alternative to effectively manage resources.

Another non-government organization, Upkar has worked for self-help group formation for *micro-financing*. Two other organization *People to People* and *Chaitanya Foundation* have also worked here for education and awareness.

In this regard, institutions have played an extensive role to fighting with water scarcity in the Banganga catchment.

6.3 Summary and Conclusions

The lesson learned from these villages has been summarized and scaled-up to the whole Aravali hill catchments (Table 6.4). It may also be concluded that if government or NGO took-up initiatives in the welfare of the people, water resources project will be sustainable.

Table 6.4 Summarized status and possible recommendation of the issues

S. no.	Issues	Status	Concluding remarks/recommendations
1	Education	Inadequate facilities	Secondary school is required in the area along with vocational training for household employment
			Encouragement for girl education is requisite for integrated development
2	Drinking water	Drinking water adequacy and water quality is all right	Practice of rain water harvesting has helped these villages for maintaining adequacy in drinking water. The lesson learned from the promotion of indigenous techniques in the Banganga catchment can be further promoted in the other catchment having similar geophysical settings
			The MNREGA scheme can be the alternative institution to support these programmes
		Ground water (GW) level is lowering	GW utilization at sustainable basis should take place in these areas. The lowering of the groundwater in the catchment is mostly because of the mining work, which should be paused or controlled. Ministry of Environment and Forests should take initiatives for unauthorized mining in the area. Mining License should be provided after cumulative impact assessment of the mining on other natural resources and livelihoods
		Rain water harvesting structures are not fully utilized because of inappropriate location and techniques	For increasing GW recharge a simple technique that could be introduced by the villagers themselves is diverting rainwater into a recharging wells or injecting wells through either an underground pipe system or through an open ditch. This can be constructed by help of local institutions and public participation
			Presently, the Johads made for GW recharge is not functioning effectively because of geological boundary condition, and tanks are working like a evaporation pan. For more recharge, groundwater recharge shaft should be installed in the tank. The recharge shaft are more efficient and cost effective technique of the GW recharge

(continued)

Table 6.4 (continued)

S. no.	Issues	Status	Concluding remarks/recommendations
3	Sanitation and hygiene	Open defecation	Community toilet may be proposed. Conversely, any development for community toilet needs proper awareness for functioning with assured water supply. Otherwise, open defecation system is appropriate as the case was
4	Electricity	Shortage of power supply	Daily supply for higher duration is required
5	Land resources	Soil is productive. In Kharif, fertilizers are less used. Crop rotation practice helped to maintain the productivity	A similar awareness regarding the environment protection needs to be developed in other parts of the Yamuna river basin
6	Forest	The forest area of village is protected. Villagers also follow rules for protecting forests	A similar awareness regarding the environment protection needs to be developed in other parts of the Yamuna river basin
7	Water Resources	Water harvesting and natural resource management is inbuilt in villages. Villagers have developed coping mechanism during dry years	Regular maintenance of structures and follow up of rules is required. However, further construction of rainwater harvesting structures needs hydrological assessment, topographic suitability, and hydrogeological investigation
			Promotion of indigenous practices should be promoted throughout the Yamuna basin depending upon the geophysical settings
8	Agriculture	Crop yields are not high	Soil conservation measure and usage of organic manure should be increased
		Cultivation cost is also high Reduced crop diversity	Crops with low water demand should be promoted. This will reduce the cost of cultivation and over exploitation of ground water
			Crop diversity should be enhanced for secure economy of village as well as better crop- water management
9	Irrigation	Dug wells source	Water quality is in the area is suitable for drinking. Water level is decreasing sharply. Unauthorized mining and licencing should be paused

(continued)

Table 6.4 (continued)

S. no.	Issues	Status	Concluding remarks/recommendations
10	Livestock	Fodder management needed	Fodder crops are grown in villages on small scale. Protected pasture area and farm forestry is required for livestock rearing
		Plantation of fodder trees	Plantation of fodder tree needs to be prioritized.
			These areas have good scope for livestock development as compared to agriculture. Livestock economy needs to be re-invented
11	Non-Farm sector	Not scaled up	Non-farm activities may include small scale or in house work for revenue generation
12	Economic security	No stabilized income Risk during aberrant weather	Crop diversity, pasture, farm forestry and non- farm activities should be implemented for stabilize income
13	Institutions	*Gram sabha*	Gram sabha in not active now. It need to be trained for supervising the development works Gram sabha may train villagers for better livestock rearing, farming and non-farm activities for revenue generation. Practice of integrated nutrient and pest management and suitable cropping pattern can also be applied through Gram sabha
			Arwari sansad is also not active. These institutions need revival for broader perspective for sustainable agro ecosystem
			Without local institutions, any development cannot be sustainable
14	Landscape	Landscape transformation	Agricultural intensification has reduced forest, pasture and open land in village. Awareness regarding impact of human activities on landscapes will serve as a useful tool in formulating regionally appropriate policies for conserving agro ecosystem. It should emphasize the impact of deforestation, crop and irrigation management, watershed treatment, and management of degraded agriculture fields, etc

References

Jain SK, Agarwal PK, Singh VP (2007) Hydrology and water resources of India. Springer, New York

Rai RK, Upadhyay A, Ojha CSP, Singh VP (2011) The Yamuna river basin: water resources and environment. Water Science and Technology, vol 66. Springer, The Netherlands

Chapter 7
Case Study: Banas Catchment

Abstract The Banas catchment has more or less similar characteristics as the Banganga catchment. However, the status of the villages in the catchment is very poor because of no involvement of government and non-government organizations (NGO). In this catchment of the Yamuna river basin immediate attention is requisite and initiatives should be taken considering the IWRM approach.

The Banas catchment is a sub-catchment of Chambal River catchment, extends over parts of Jaipur, Dausa, Ajmer, Tonk, Bundi, Sawai Madhopur, Udaipur, Rajsamand, Pali, Bhilwara and Chittorgarh districts. The total catchment is comprised from area of 45,833 km^2. River Banas originates in the Khamnor hills of the Aravali range (about 5 km from Kumbhalgarh) and flows along its entire length through Rajasthan and culminates in river Chambal near village Rameshwar of Khandar block of Sawai Madhopur district (Jain et al. 2007; Rai et al. 2011). The total length of the river is about 512 km. The Important tributaries of Banas are Berach and Menali that joins at right bank, and Kothari, Khari, Dai, Dheel, Sohadara, Morel and Kalisil on the left bank. In Banas catchment there are 7 major (Rajsamand, Meja, Chhaparwara, Kalakh Sagar, Tordi Sagar, Morel Bund and Galwa Bund), 33 medium and 1,219 minor irrigation projects have been constructed. Apart from these several small irrigation systems (covering less than 20 ha) have been constructed and operated by *panchayat samities*. Gosunda and Bisalpur Irrigation Project are on river Banas. Hydrogeologically, Banas catchment is similar to the Banganga catchment.

Primary survey and campaigning was conducted in Bikaran and its surrounding villages in Bundi district of Rajasthan in Banas catchment.

A. Upadhyay and R. K. Rai, *Water Management and Public Participation*, SpringerBriefs in Earth Sciences, DOI: 10.1007/978-94-007-5709-7_7,

7.1 Geo-Physical Characteristics

7.1.1 Climate and Rainfall

Area is rainfed with moderate climate. The erratic and low rainfall (ranges between 200 and 400 mm) during the last 5 years induced water related problems. The annual rainfall in the catchment varies between 450 and 600 mm with mean of 541 mm. Most of the rainfall (i.e. 95 %) experienced during the four monsoon months (June–September).

7.1.2 Topography and Soil

Orographically, the western part of the catchment is marked by hilly terrain belonging to the Aravali chain. East of the hills lies an alluvial plain with a gentle eastward slope. Surface elevations in the western hilly part ranged between 850 and 1,123 m, while in the alluvial plain it varies between 580 and 750 m. Topography is undulating with small hills in the nearby areas.

Soil is red–yellow with 1–3 ft in depth. Terracing and bunding will help here for soil conservation. Slightly shallow soil depth is also encountered in few areas (6–12 in). Soil texture in the area can be categorized as sandy loam.

7.1.3 Biodiversity

Among trees Boswellia serrata, Angoeissus pendula, Acacia catechu, Manilkara hexandra, Acacia nilotica, Azadirachta indica, Calotropis procera are commonly found in this area. Among fauna, wolf, jackal, ape, and neelgai are common species in this area as informed by villagers.

Common crops: Maize (Makka), fodder (Jwar-chari), Til, Soyabean during Kharif and Wheat, Mustard (Sarson) and Masur during rabi season are grown.

7.2 Inferences from Primary Survey

7.2.1 Water Conservation

In the catchment, various indegenous water harvesting system have been installed. Bikaran village has a small *anicut* which was constructed in year 2003 (Fig. 7.1). Villagers informed that *anicut* has helped the villagers to survive as they had some

Fig. 7.1 Anicut at Bikaran, Bundi

production even in previous drought year. The erratic and low rainfall just in the range of 200–400 mm during past 4 years did not affected the village, while the situation in surrounding area was worst because of no water conservation measures have been taken up. The comprehensive water resource development and management through various innovative technologies and local wisdom have successfully regenerated the vegetation and improved the surface and ground water resources. It was also noted that villagers have adopted the soil and water conservation techniques. In the undulating terrain, *Medbandi* has been practiced in agriculture fields to avoid soil erosion due to water and wind. However, rainfall during the year 2008–2009 were not helpful in filling the anicut, but certain moisture retained helped them in taking one crop due to *medbandi*. Villagers were able to harvest agricultural production, and have water for drinking in summer, fodder, surplus milk to sale. People developed the courage and skills to manage resources for their own well being as well as environmental protection.

Construction of waste weir, *nadi* and *anicut*, and maintenance of old water harvesting structure has been done under schemes of MNREGA in villages of Pench ki Bawri. Village Bhimganj, Motipura, Paprala, Pench ki bawri, Bhil ka Khera and Kalamal have been benefited from implementation of such scheme. Road and small bridge has also been constructed in Bikaran under this project. A major water harvesting structure surrounding the Bikaran village of the Bundi district is shown in Fig. 7.2.

7.2.2 Local Institution

Villagers know the importance of water conservation and have constructed *anicut* and *medbandi*, etc. They do not have any formal or informal institution so that they can get some training and awareness for further improvement in the soil and water

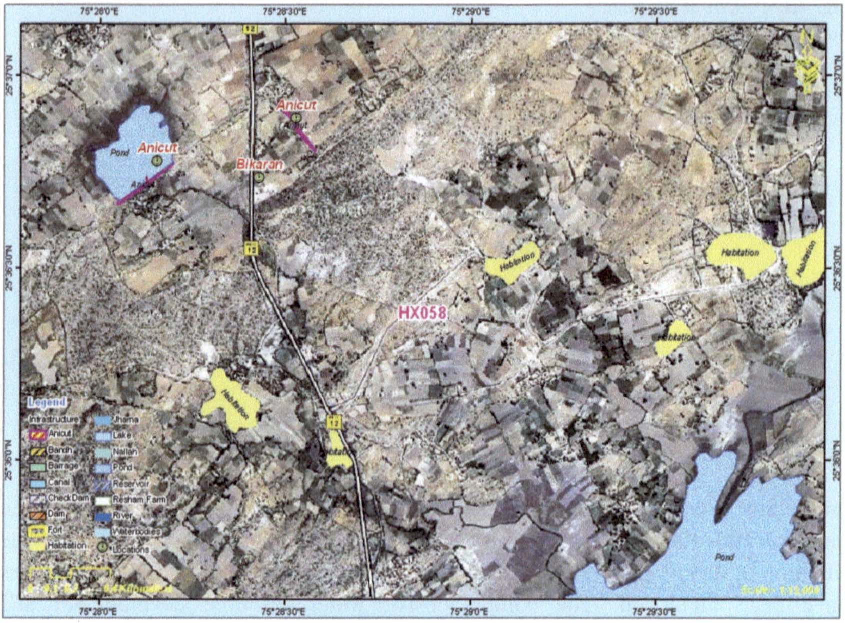

Fig. 7.2 Major WHS and habitation in Bikaran

conservation techniques. Villagers showed assertiveness for need of training and awareness for better crop production and water conservation. However, at the time of construction of *anicut*, water user association was formed. But it was never functional.

7.2.3 Environmental Status

Drinking Water Adequacy and Quality: Drinking water availability is adequate in Bikaran. Water is potable. Villagers informed that the construction of *anicut* and *medbandi* work has improved the water availability in the wells (Table 7.1). However, in the recent years, declination in the water table has been reported because of scanty rainfall which was insufficient to fulfil the *anicuts*.

Table 7.1 Water level in open well in Bikaran (m bgl)

Year 1990		Year 2005		Year 2010
Pre monsoon	Post monsoon	Pre monsoon	Post monsoon	Pre monsoon
30	25	27	18	28

Source Primary survey

In totality, twenty one wells exists in the Bikaran village. Centrifugal pump installed in 17 wells run by diesel engines are generally used. Only one electric operated centrifugal pump exists in the area. Two hand pumps are also exists for drinking water.

Sanitation and hygiene: Open defecation is practiced in village. Area comprising hills and hillock which supports hygiene. With the support of Government of India and UNICEF, the rural sanitation programme was run in nearby areas during 1995–1996. Total sanitation campaign was also launched in about 10 districts in 2001 including Bundi. This programme was initially being implemented in the field by the Panchayati Raj Department through Zila Parishad, Panchayat Samities etc., and it has now been transferred to Public Health and Engineering Department (PHED).

Waste Disposal: Waste is disposed in low land area and barricading by bush has been done, called as *Dhalau.*

Disease: As such diseases are not prevalent in the village. Area is dry most of the time and nature takes care.

Fuel: Fuel wood and cow dung are mainly used for cooking. Women collect dry bushes from nearby hills for fuel wood.

7.2.4 Socio-Economic Status

Population and landholding: Bikaran village has about 480 populations with 50 household. Villagers have marginal to small operational land holding with 1–2 *bigha* (1.75 *bigha* equals 1 acre). Only 3–4 household have 10–15 *bigha* land. About 10 % villagers come under landless category.

Education: Nearly 40 % population is literate. Bikaran has a primary school. Villagers generally send their children to school.

Occupation: Occupation can be categorized as farmers, agricultural labourers and masonry workers. Women also work as labourer and help in agriculture they have also participated in MNREGA. In the village, 70 % population of working class is labourer.

Migration: Many villagers have migrated to Kota, Jhansi and Jaipur for mining and masonary work.

Status of women: Women do household work, drinking water collection, agriculture and livestock rearing. Men work as labourer in nearby towns, causing more responsibility of work to women. Women face problem in getting fuel wood as village forests are degraded and nearby hills come under Forest Department jurisdiction. It was informed that tree lopping for livestock is not available and a cattle rearing is very costly, for which women devote time for livestock grazing.

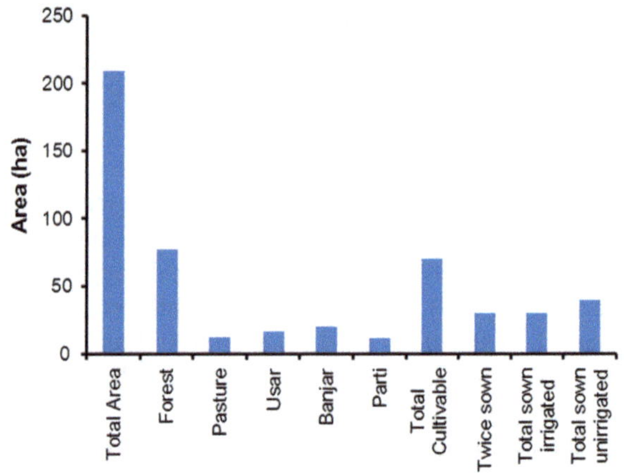

Fig. 7.3 Land use in Bikaran village (2008–2009)

7.2.5 Agriculture and Irrigation

Total area of Bikaran is 209 ha of which 77 ha is forest land and 70 ha total cultivable area (Fig. 7.3). Major crops are *Makka, Til, Chana, Gwar, Wheat, Sarson, Mirchi, Bhindi, Pyaj and Matar*. In recent 2 years, agricultural production has been reduced. Impact of Gudha dam[1] has considerable impact on the agriculture. When this dam is full, land receive good soil moisture for crop production. *Kharif* and *rabi* crops are taken mainly in Bikaran. Area under *rabi* crops are irrigated, whereas, kharif crops are cultivated as rainfed. Figures 7.4 and 7.5 shows the cropped area under *rabi* and *kharif*, respectively.

Crop Production: Villagers informed that crop production has increased due to availability of water because of rainwater harvesting structure (Table 7.2). However, production of maize and wheat has been increased due to applications of high yield variety.

Fertilizer and Pesticide Application: Farmers generally depends on chemical fertilizers, though the organic manure is also used depending on the availability as they have only few cattle. Villagers informed that fertilizer consumption has almost doubled due to lack of animals. The fertilizer application has been increased by 130 kg/ha in the recent years, which was only 60 kg/ha 10–15 years back.

Irrigation: Tubewells operated by diesel engine are generally used for irrigation. In *Kharif* all crops are un-irrigated except vegetables. *Rabi* comprises all irrigated crops. However, *masur* in 1 ha and *sarson* in 2 ha was un-irrigated.

Livestock: Very few animals are reared by villagers (Table 7.3). Villagers are labourers in nearby areas and prefer masonry work than farming and rearing cattle. Dairy farming is a reliable source of livelihood for small and marginal farmers.

[1] Gudha dam is among one of the largest earthen dam of Asia.

Fig. 7.4 Cropped area during kharif in Bikaran (2008–2009)

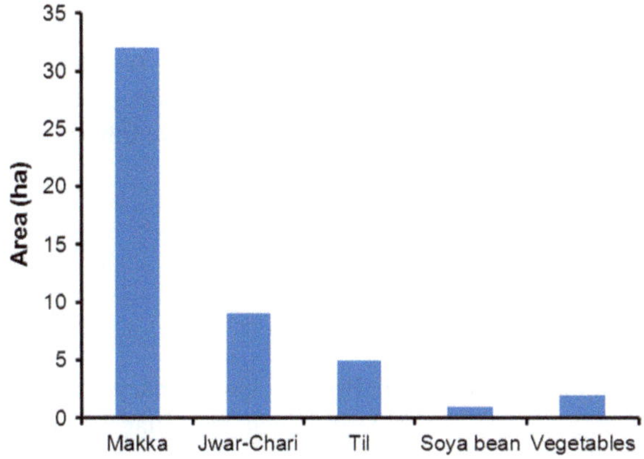

Fig. 7.5 Cropped area during *rabi* in Bikaran (2008–2009)

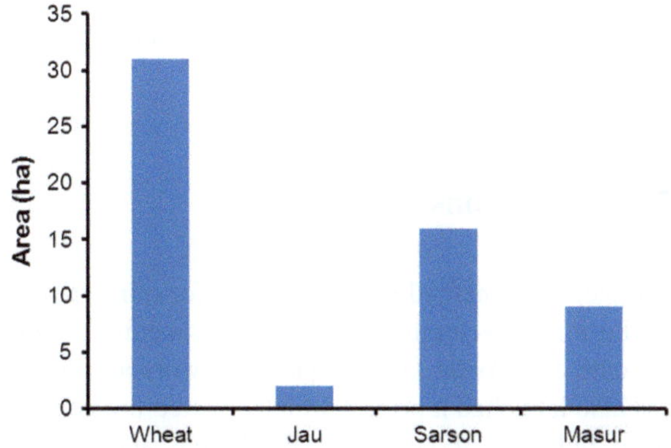

Table 7.2 Crop production in Bikaran (kg/ha)

Crop	Year 2000	Year 2005	Year 2008
Wheat	1,800	2,300	2,350
Makka	560	800	850
Jau	1,500	1,400	2,000
Sarson	680	750	800

Source Primary survey

Table 7.3 Livestock population in Bikaran

Livestock	Year 1990	Year 2010
Cow	25	20
Buffalo	40	35
Bullock	10	10
Goat and Sheep	1,000	500

Source Primary survey

However, lack of pasture land and insufficient rainfall are major constraint. It is therefore, villagers are forced to develop revenue from non-farm activities.

Farm Power and Machinery: Villagers borrow bullock and tractor from other village for ploughing. However, they have other regular appliance used in farming such as plough etc.

7.2.6 Amenities

School and Hospital: Bikaran has primary level government school. Swasthya Kendra is also available in Panchayat village. For medical facility villagers go to Hindoli.

Road: Bikaran is well connected to pacca road and National Highway.

Electricity: Bikaran has electricity supply; however, regular power failure is common.

7.3 Institutions

Village was benefited with Panchayat for rainwater harvesting structure. However, institution development needs reawakening. *Panchayat Samity*, MNREGA Scheme and Forest Department have worked effectively in these areas. UNICEF has also launched a project which covered the nearby villages. NGO like Bhartiya Agro Industries Foundation (BAIF) has also worked in the area for watershed development and agricultural promotion.

In most of the government and non-government schemes, for overall socio-economic development in the area, public was involved with high passion. However, in few villages, where government has not taken any initiatives or local institutions are non-functional, public participation is absent in the development work because of financial and technical constraints.

7.3.1 Capacity Building

The construction and maintenance of rainwater harvesting structure needs proper training to water user association. There is need to revive WUA so that watershed can be managed within the IWRM framework. The investment for rain water harvesting has already been made and people are also appreciating as they have directly benefited. The benefits may be enhanced through capacity building of WUA so that ignored important components (viz., small check dam for soil conservation, pasture development and plantation in nearby areas) can be adequately considered.

Table 7.4 Summarized status and possible recommendation of the issues

S. No.	Issues	Status	Concluding remarks/Recommendations
1	Education	Inadequate amenities for education	Higher schools as well as vocational training centres are required
2	Drinking water	Adequate	Drinking water is adequate in all respect
		Ground water (GW) level is lowering	GW utilization at sustainable basis should take place in this area
			For increasing GW recharge simple technique could be further encouraged based on the lesson learned from the Banganga catchment
			In this area also groundwater recharge shaft should be promoted with the help of state ground water department
3	Sanitation and hygiene	Open defecation	Community toilet may be proposed
			Open defecation system is appropriate in the region because of climate and topographic characteristics of the catchment
4	Electricity	Electricity supply is uncertain	Daily supply should be maintained
5	Land resources	Severe soil erosion on high sloped terrain	Soil conservation measure can be taken up with the help of forest department. Biological as well as structural measures can be taken up to protect the soil. Gully plugs could be constructed on upstream of the check dam, which will slow down water and will help in GW recharge along with reducing silt load into the check dams

(continued)

Table 7.4 (continued)

S. No.	Issues	Status	Concluding remarks/Recommendations
6	Forest	Deforestation	Villages should have forest and pasture for livestock grazing and other by products for revenue generation Protected area should be developed for limited use so that the regeneration capacity of the forests is achieved. Forest department may intervene Because of scanty rainfall, green vegetation has been scarced in the area of shallow soil depth. Biological and structural measures can help in afforestation
7	Water resources	Need of more small rain harvesting structure	Water harvesting structures and soil conservation structures need to be constructed Water resources management programme can be initiated, which include land use planning, crop planning, water resources allocation, irrigation management, etc
8	Agriculture	Low crop yields	Soil conservation measure and usage of organic manure should be increased
		Increasing cultivation cost Low crop diversity	Crops having low water demand should be encouraged. This will reduce the cost of cultivation and over exploitation than recharge Crop diversity should be enhanced for secure economy of village as well as better crop- water management Mixed farming should be encouraged for sustainable agriculture

(continued)

Table 7.4 (continued)

S. No.	Issues	Status	Concluding remarks/Recommendations
9	Irrigation	Dug wells	Water quality of water is satisfactory; however, extraction cost is very high
10	Livestock	Very few in numbers	Livestock economy needs to be developed. Plantation of fodder tree needs to be prioritized
			Protected pastures and farm forestry land may be developed. This will also reduce the drudgery faced by women for fuel wood collection
11	Non-Farm sector	Not scaled up	Non-farm activities may include local craft work, embroidery etc. Women may be encouraged to work for these for revenue generation
12	Economic security	No stabilized income Risk during aberrant weather	Crop diversity, pasture, farm forestry and non- farm activities should be implemented for stabilize income
13	Institutions	Water user association non functional	Water user association needs to be revived. Capacity building and training for better cropping and maintenance and up keep of the water harvesting structure
14	Landscape	Landscape transformation	Forests are declining. Village level institution can work for protecting nearby forests. Awareness about the changing effects of landscapes with reducing forests, pressure on agriculture and irrigation can serve as a useful tool in capacity building at village level

7.4 Summary and Conclusions

In general Banas catchment has very undulating topography and scanty rainfall, therefore to enhance the socio-economic growth; integrated water resources management will be more effective and benefiting. Public are aware about their role in the future development but need extensive support from the government. The quantitative and qualitative analysis helped in formulating the recommendation for water management in the Banas River catchment (Table 7.4). These recommendations will be very useful in formulating a new water related projects of similar catchments, following summarized.

References

Jain SK, Agarwal PK, Singh VP (2007) Hydrology and water resources of India. Springer, New York

Rai RK, Upadhyay A, Ojha CSP, Singh VP (2011) The Yamuna river basin: water resources and environment. Water Science and Technology, vol 66. Springer, The Netherlands

Chapter 8
Case Study: Chambal Catchment

Abstract The village selected in the catchment has severely affected by the water scarcity, though the villages are surrounded by two perennial rivers, namely the Chambal and Alnia River, a tributary of Chambal. Because of water scarcity, public are migrating to the nearby area where they can get water for their animals. In the area, both soil and water conservation practices should be initiated at priority basis. Support from the Government and local institution or water related NGO is completely absent and therefore, immediate attention is required.

River Chambal, a principal tributary of river Yamuna, originates in the Vindhyan ranges near MHOW in Indore District of Madhya Pradesh. The river flows for some 320 km in a generally northerly direction before entering a deep gorge in Rajasthan at Chourasigarh, about 96 km upstream of Kota. The deep gorge extends up to Kota and the river then flows for about 226 km in Rajasthan in a north-easterly direction, and then forms the boundary between MP and Rajasthan for about 252 km. Thereafter, the river forms the boundary between Madhya Pradesh and Uttar Pradesh for about 117 km, enters Uttar Pradesh near Charak Nagar village and flows for about 40 km before joining river Yamuna. The Banas and Mahi River catchment lie to its west, and Gambhir and Parbati river catchments to its north. Its eastern and southern edges border Madhya Pradesh State. The total drainage area of the Chambal catchment is 1,43,219 sq km. Orographically, the western part of the Basin is marked by hilly terrain belonging to the Vindhyan chain, with fairly sloping terrain along the Chambal River and its tributaries. East of Kota-Jhalawar lies an extensive alluvial plain sloping gently northwards towards the Yamuna River in Uttar Pradesh. The mean annual rainfall over the Chambal Basin was computed as 797 mm, of which about 93 % falls during the four Monsoon months (June–September). In the catchment, there are 7 Major (Gandhi Sagar, Rana Pratap Sagar, Jawahar Sagar, Kota Barrage, Parwati Pick-up Weir, Harish Chander Sagar and

A. Upadhyay and R. K. Rai, *Water Management and Public Participation*, SpringerBriefs in Earth Sciences, DOI: 10.1007/978-94-007-5709-7_8,

Fig. 8.1 Dharampura pond, Kota

Fig. 8.2 Dharampura village near the pond, Kota

Gudha Dam), 12 Medium and 134 Minor irrigation projects in Chambal River Basin, as well as some small irrigation systems (covering less than 20 ha) constructed and operated by Panchayat Samities (PS) (Rai et al. 2011).

In the catchment, two villages Dharampura and Pachankui of the Ladpura block of Kota, which lies between river Chambal and Kali Sindh has been identified for the survey and investigation. The nearest dam i.e. Alniya dam on Alnia river is very close to these villages. Ladpura comes under Chambal canal command. Diversion channel is being constructed for diverting water to Alniya for flood protection in the Kota. In spite of rich water resources in the catchment, these villages have worst status in terms of drinking and irrigation water (Figs. 8.1, 8.2, 8.3). A waste weir at Bandha Dharampura village has been constructed on *nallah* to prevent southern area (such as Vigyan Nagar, Talwandi, Mahavir Nagar etc.) of Kota city from flooding. A drainage channel of 5.60 km long up to Chambal River for diversion has been taken up, 75 % work has been completed.

Fig. 8.3 Alniya dam, Kota

8.1 Geo-Physical Characteristics

8.1.1 Climate and Rainfall

The area comes under a sub-tropical climate with moderate to hot temperature, ranges between 7 and 45 °C, and relative humidity drops to about 20 % during March, April and May. July–August is most humid period. During this period humidity goes to 70–80 %. The Kharif-wet season is spread from July to September. Rabi-winter season is extended between the months of October and February, while the months between April and May is pre monsoon season which is usually hot. The monsoon usually starts from second half of the June but heavy rainfall experienced during July to August and early September. The mean annual rainfall over the area is 850 mm.

8.1.2 Topography and Soil

The soils of Chambal command area are alluvial in nature and fairly deep in valley portions. However, the depth varies from place to place depending upon the extent of weathering, natural slope, and drainage, extent of erosion and exposure of rock out crops. The slope generally ranges from flat to 2 %. These are dominantly fine textured soil varying from clay loam to clayey with mostly clay loam. These are non-calcareous to slightly calcareous geological formation in nature.

8.1.3 Biodiversity

Acacia nilotica, Acacia catechu, Prosopis cineraria, Azadirachta indica, Zizyphus mauritiana, Z.nummularia, Dalbergia sissoo, Ficus religiosa, Madhuca longifolia

Cassia fistula, Acacia leucophloea, Pithecellobium dulce, Boswellia serrata Calotropis procera and *Argemone maxicana* were seen in the area. Due to deforestation and lack of conservation practice forests have been wiped; only counted numbers of trees are available. However, the species present shows that earlier the area was not like this. This area is more suited to pasture and farm forestry development and needs land treatment. Through protection and plantation the forests can be rejuvenated. As the forests are over, fauna is also lost. Villagers informed that wolf, *neel gai*, deer, wild boar are common wild species in this area.

8.1.4 Common Crops

Maize, Fodder-Jwar, Til, Soya bean, Paddy, Rose flower in kharif and Wheat, Chana, Jau and Sarson in Rabi are main crops in this area.

8.2 Inferences from Primary Survey

8.2.1 Water Conservation

Dharampura has one large pond (10.48 ha) which needs rejuvenation (Figs. 8.4 and 8.5). At the time of visit very little water was available in Dharampura pond. The flood protection work for Kota through diversion canal system has further accelerated the water scarcity problem in the pond. Rain water harvesting and soil conservation steps have not been undertaken in these villages. Under MNREGA scheme little work has been proposed for Pachankui and Dharampura. Villagers informed that as their village come under *Nagar Nigam* so they are not able to avail the *Panchayat* facility. Pachankui also has a pond (Fig. 8.6). Water availability is worst in Pachancui. These villages have developed since last 40 years back. These people migrated here with their livestock because of pasture and forests were available. Soon they started cultivation; however, the soil depth is very low. They spread soil on their land to increase soil depth suitable to cultivation. At the time of visit, soil spreading in agriculture fields was going on. Any transformation of landscape needs proper conservation practice in the absence of it, it fails. However, the problem can be restored through suitable soil and water conservation practices. Villagers informed that this terrible situation started from last 6 to 7 years before this problem was not so acute. They have barricaded fields through local stones so that soil erosion can be checked and water can be retained in field. Soil depth is very low it varies 0–4 inch in general and not more than 12 inch in any field.

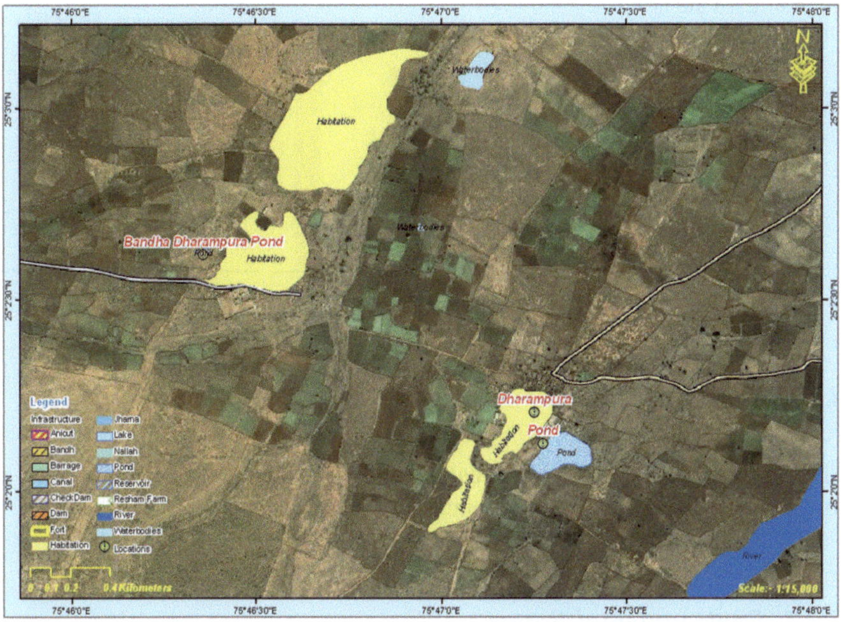

Fig. 8.4 Major WHS and habitation in Dharampura village

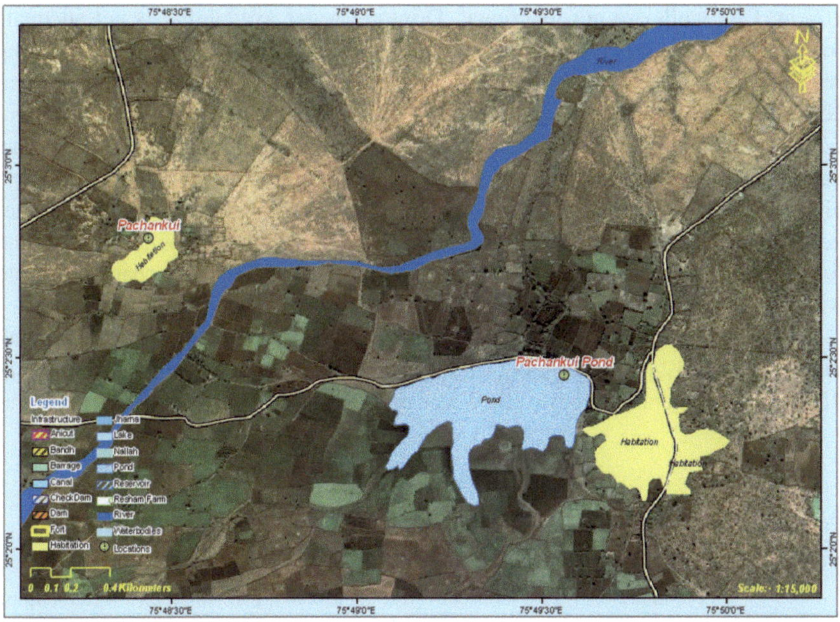

Fig. 8.5 Major WHS and habitation in Pachankui village

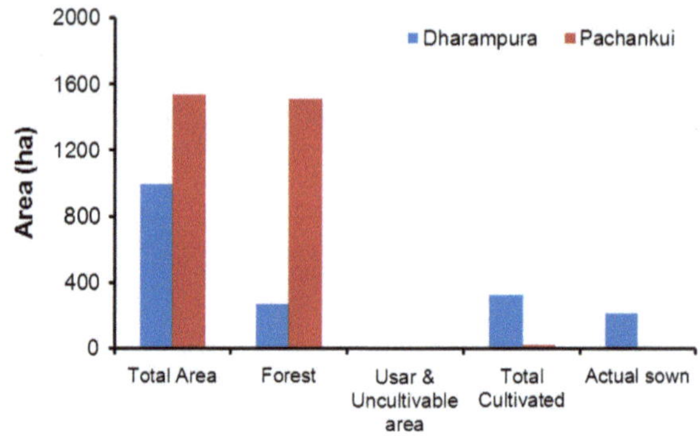

Fig. 8.6 Land use in Dharampura and Pachankui (2008–2009)

8.2.2 Local Institution

These villages do not have any formal or non-formal institution. In Dharampura at low scale self-help group is functioning for *micro financing*.

8.2.3 Environmental Status

Drinking water adequacy and quality: Ground water table in the area is at 90–107 m. Dug wells are not functional. Drinking water availability is not adequate for human beings as well as for animals. People have to go 250 m for drinking water collection and household work. Village has 25 wells and only one has water. Nine wells, which have been installed in the natural drainage, are also not functional. These well are 7.5–10.0 m deep. Villagers informed that water storage augmentation work for Alnia dam has distorted the old storage system. However, villagers did not tried for deepening of these well as they have tube well for water need. In Pachankui village drinking water facility is fulfilled by hand pump and tube well. Water quality is satisfactory in available sources. Pachankui has no working dug well at present (Table 8.1).

Dharampura has 11 hand pumps out of which only one was functional. Pachankui has 12 hand pump out of these 4 are working. Dharampura village has 22 electricity operated tubes well. Water is available at 90–105 m. Pachankui has

Table 8.1 Depth of open well in Dharampura and Pachankui (m bgl)

Village	Year 1985		Year 2000		Year 2010
	Pre monsoon	Post monsoon	Pre monsoon	Post monsoon	Pre monsoon
Dharampura	30	20	No water		No water
Pachankui	35	27	No water		No water

Source Primary survey

5 tube well operated by electricity. Tube well is source for drinking water and irrigation in both the villages.

Sanitation and hygiene: Open defecation is common practice in both the villages.

Solid waste disposal: Open dumping is done for waste disposal. Agricultural waste is used for making manure to use in agricultural fields.

Disease: No waterborne communicable disease, pathogenic-or non-pathogenic is prevalent in this area. Women informed problem of dizziness and suffocation due to cow dung and fuel wood use in cooking.

8.2.4 Socio-Economic Status

Population and landholding: Dharampura village has about 950 populations with 100 household. Gurjar, Regar and Bhil (tribal) reside here. About 90 % population comes under landless category. Only 10 % household has 1–2 bigha land. Pachankui village has population of about 650 with 80 household comprising Bhil, Regar and Gurjar as dominating casts.

Education: In Dharampura most of the villagers are illiterate. Literacy is less than 10 %. Village has government school up to 8th class. Previously, students from outside of villages enrolled, however, students of Dharampura have also enrolled in recent years. Staff of school is very cooperative and they are motivating the villagers for uplifting through education and natural resource conservation. Pachankui has a government school up to middle.

Occupation: Shepherds mainly reside in these villages. They prefer to migrate instead of practicing conservation methods. In Dharampura, people depend mainly on rearing animals. Nature is harsh and because of non-availability of pasture and water, people are moving to nearby areas where they can feed their animals. About 75 % of villagers migrate with animals in dry months. People work in sand stone mines and stone crushers. Villagers in Pachankui are labourers in the nearby cities and work in mines and masonry work, a few are farmers and shepherds.

Migration: Migration is very high almost each family has person migrated for work as labourer. Villagers informed that if they get irrigation water and soil conservation measures they will be able to survive in the village. These villages look like no man land as almost all workable go for labourer job, because of adversity of nature and lack of conservation measures. In Pachankui, status is inferior as compared to Dharampura even.

Status of women: Women collect drinking water and fuel wood. There is no proper medical facility. As these village are in remote area of Ladpura conveyance to come in city is also not available at time of any emergency. During discussion with women they showed interest for craft and embroidery work. Women showed their interest for training and awareness for health and hygiene.

Fuel: Fuel wood and cow dung is used for cooking. Kerosene oil is also used by few families sometimes. In Dharampura, only 2–3 household have LPG stoves.

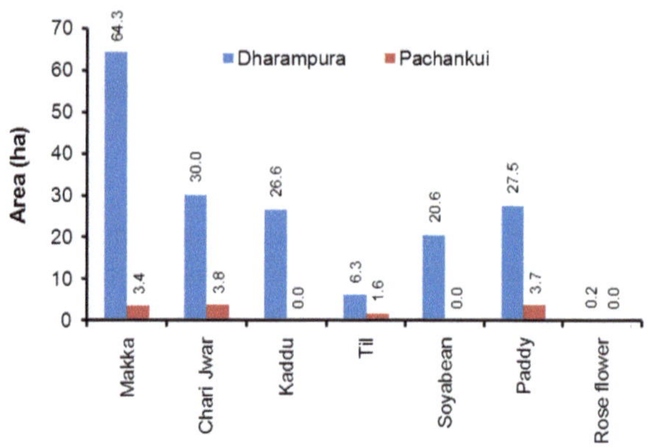

Fig. 8.7 Kharif crop in Dharampura (2008–2009) and Pachankui (2009–2010)

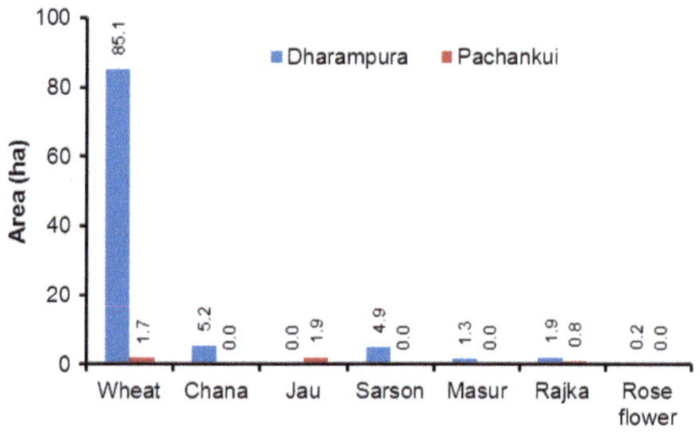

Fig. 8.8 Rabi crop in Dharampura (2008–2009) and Pachankui village (2009–2010)

8.2.5 Agriculture and Irrigation

Agriculture: Dharampura has forest area of about 27 % of its total geographic area of 993.18 ha (Fig. 8.6). Only 21 % area is under cultivation. Agriculture is poor in Dharampura, very few fields were seen under cultivation in the village because of water scarcity. Most of the area is extended by the underneath impermeable formation with shallow soil depth. In Pachankui, 98 % of the area is under forest category in land use. It has only 0.8 % actual sown area. The crop diversity is very poor. They harvest 3 crops in *kharif* and 3 in *rabi* season (Figs 8.7, 8.8). Villagers informed that they are not able to store grains for seed. They buy it from local market.

Crop Production: Dharampura and Pachankui have lower production of crop as compared to other cases studied. Year 2008 was best crop year as informed by villagers due to good rainfall (Table 8.2).

Irrigation: Tube wells are used for irrigation in both the villages. During *rabi* season all the crops are irrigated. In *kharif* season rose flower is the only irrigated crop.

Livestock: In both the villages, livestock population has been considerably reduced (Table 8.3). Buffalo, cow and goat are the main livestock. Few families also have sheep.

Table 8.2 Crop production in the selected villages (kg/ha)

Crop	Year 2000		Year 2005		Year 2008	
	Dharampura	Pachankui	Dharampura	Pachankui	Dharampura	Pachankui
Wheat	1,700	1,600	1,780	1,650	2,200	2,100
Maize	400	350	450	400	500	430
Gram	500	400	550	430	600	600
Mustard	500	500	600	550	700	700

Source Primary survey

Table 8.3 Animal population in Dharampura and Pachankui

Livestock	Year 1990	Year 2010
Dharampura		
Cow	80	40
Buffalo	120	90
Goat & Sheep	3,000	2,000
Pachankui		
Cow	60	50
Buffalo	80	60
Goat & Sheep	4,000	3,000

Source Primary survey

Farm Power and Machinery: Villagers own small appliances for agriculture. However, for ploughing they borrow bullock as well as plough from fellow villages.

8.2.6 Amenities

Infrastructure: Both the selected villages have school up to middle class. Swasthya Kendra and Anganwadi are functional. Dharampura has capacity for developing religious tourism. Both the villages are connected to *pacca* road. However, road quality is poor, may be because of construction going on for canal diversion, flyover on National highway and Sainik Pravas Samity.

Electricity: Electricity supply is unavailable most of the time.

8.3 Institution

Irrigation Department has intervened in the area; however, this has actually deprived them in water. There is no existence of formal/non-formal institution which can carry out capacity building. MNREGA can also help this area, but for the limited time. Village level small institution like WUA or G*ramsabha* is required, which will give platform for discussion and awareness.

Table 8.4 Summarized status and possible recommendation of the issues

S. No.	Item	Status	Concluding remarks/Recommendations
1	Education	Inadequate amenities	Schools as well as vocational training centers are required
2	Drinking water	Drinking water scarcity Ground water (GW) level is lowering	Secure drinking water supply is required. Abandoned dug wells may be cleaned and deepened for usage Water supply should be maintained through surface water resources because of unfavorable geological formation
3	Sanitation & hygiene	Open defecation	Community toilet may be proposed. It needs awareness before application of such interventions with assured water supply Climate, topography and population density supports the open defecation system. Sewerage system will create water crises and environmental problem
4	Electricity	Frequent power cuts	Daily supply need to be maintained
5	Land resources	Severe soil erosion on slopes and along streams. Due to erosion, and sedimentation, capacity of the ponds has been considerably reduced	Conservation measures are required at various locations in Dharampura and Pachankui Soil and water conservation measures should be initiated on priority basis
6	Forest	Deforestation Lack of water and food led the wild animals destroy surrounding croplands	Deforestation has lost the whole forest cover. Villages should have forest and pasture for livestock grazing and other by products for revenue generation Protected area should be developed for limited use so that the regeneration capacity of the forests is achieved Plantation of fodder and fruit trees required

(continued)

Table 8.4 (continued)

S. No.	Item	Status	Concluding remarks/Recommendations
7	Water resources	High surface runoff during Monsoon	As the area has undulating terrain and rocky, the runoff rate is very high during rains. This leads to soil erosion and loss of water from the catchment
		No rain water harvesting structures except the old pond/talai	Water harvesting structures and soil conservation structures will be helpful to rejuvenate the system
8	Agriculture	Low crop yields Low crop diversity	Soil conservation measure and use of organic manure should be increased
			Crops with low water demand should be promoted to reduce the cost of cultivation and over exploitation of GW than recharge
			Crop diversity should be enhanced to reduce the economic risk farmers as well as better crop-water management
9	Irrigation	Tube well	Tube wells are used for irrigation. Water quality is all right, however, extraction of water is very high. Rain water harvesting is required
10	Livestock	Lack of Fodder management	Integrated water resource management needs to be implemented at watershed level
			Plantation and regeneration of fodder tree needs to be prioritized
			Protected pastures and farm forestry land may be developed as these villages have tremendous scope for these activities
11	Non-farm sector	Not scaled up	Non-farm activities handicraft, embroidery etc. may be intervened. Women may be encouraged to work for these for revenue generation

(continued)

Table 8.4 (continued)

S. No.	Item	Status	Concluding remarks/Recommendations
12	Economic security	No stabilized income Risk during aberrant weather	Afforestation, pasture, farm forestry, crop diversity, and non- farm activities should be implemented to stabilize income. As the area is at high risk of erosion, damage of crop fields needs soil conservation measures
13	Institutions	No formal/informal institution exist	Villagers need training for water harvesting for which institutional development and capacity building is required Soil conservation work through community participation is needed
14	Landscape	Landscape transformation	Forests are declined. Institutions should work effectively for sustaining forests in association with development. Awareness about the effects of deforestation

8.3.1 Capacity Building

Villagers need awareness about water harvesting and watershed management and its impacts on sustainable growth. Department of Forest and Watershed Management, and Soil Conservation Department may intervene for improving the status. Public participation in fighting with the water crisis was absent in the area because of lack of technical knowledge and financial support and most important the absence of local institutions.

8.4 Summary and Conclusions

This part of the catchment has extreme water scarcity because of its geological settings. Only surface water resource augmentation is the solution for this region. However, sustainable socio-economic development requires much attention towards the soil conservation. Since cultivable area in the region is comparatively less, therefore livestock growth or other self employment schemes need to be promoted by the government. The recommendations (Table 8.4) arrived at will be very useful to the decision makers and water resources planners for sustainable development.

Reference

Rai RK, Upadhyay A, Ojha CSP, Singh VP (2011) The Yamuna river basin: water resources and environment. Water Science and Technology, vol 66. Springer, The Netherlands

Chapter 9
Case Study: Sind and Betwa Catchments

Abstract In the present case study, the villages selected are close to the river Betwa, Sind and Dhasan, where many reservoirs for irrigation and hydropower development project has been developed. In spite of water bodies surrounding the villages or area, the water problem for the livelihoods and socio-economic development remained same. Based on the investigation, land and water are not optimally managed, and the problem can be solved if IWRM approach is adopted. In the area, public participation is high but government biasedness on the basis of religion and cast has damaged the public cooperation.

The Sind and Betwa rivers originate from the state of Madhya Pradesh at Bhopal and Vidisha, respectively. Both the origins are very close to each other, and generally travel across the Bundelkhand region or Vindhyan range of the Madhya Pradesh and Uttar Pradesh. The hydrogeological characteristics of both the catchments are more or less similar. In spite of good mean annual rainfall, both the catchments experiencing frequent drought because of geological and topographical characteristics. The catchments have very undulating terrain and have good hydro power potential because of head. The river Sindh originates on the Malwa Plateau in Vidisha district, and flows southwest through Madhya Pradesh to join the Yamuna in Etawah. River Betwa originates from the Vindhyan ranges in Bhopal district at an elevation of around 390 m above sea level and meets the Yamuna in Hamirpur district. The Betwa River largely flows through rocky terrain with steep gradients; many sites are suitable for hydel power stations. An important tributary of the Betwa is the Dhasan, which also rises in Bhopal district. It then flows through Sagar district and forms the southern boundary of Lalitpur district. It joins the river Betwa in Jhansi district (Jain et al. 2007; Rai et al. 2011).

In the study two districts of Bundelkhand region i.e. Jhansi of Uttar Pradesh and Tickemgarh of Madhya Pradesh have been studied. In Jhansi, Ghisoli and Raksa villages of Babina block were selected for study, in which Ghisoli village falls under river Betwa catchment, whereas, Raksa comes under river Sind catchment of

A. Upadhyay and R. K. Rai, *Water Management and Public Participation*, SpringerBriefs in Earth Sciences, DOI: 10.1007/978-94-007-5709-7_9,

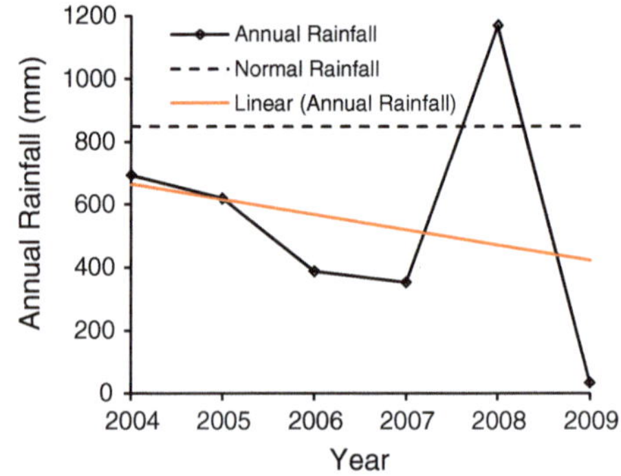

Fig. 9.1 Annual rainfall trend in the Jhansi region

the Yamuna river basin. The Tikamgarh district is famous for Chandeli ponds; these have been able to provide drinking water and irrigation facilities from the medieval days. But at many part these are being encroached for cultivation in villages and in the towns. These ponds have been serving the most important purpose of collecting the rain-water and recharging the ground water so that the drinking water well remains filled with water throughout the year. Azadpura and Kumharra village of Tikamgarh were selected for the study. In Kumharra village a Baoli (step well) is still working since last 100 years or more. The Azadpura village was earlier known as Dhimarpura, Azadpura name is given to this village because the legend Shri Chandrasekhar Azad, a freedom fighter lived here as Saint for 1.5 years in exile after Kakori case. He built an ashram and practiced shooting in forest.

9.1 Geophysical Characteristics

9.1.1 Climate and Rainfall

The distribution of rainfall is erratic. Maximum and minimum temperature varies in the range of 3–4° and 43–47°C, respectively. The May and June are the hottest months and sometime temperature goes up to 48.0 °C and minimum temperature falls below 2.0 °C during December–January. The annual rainfall has large spatial variation in the catchments. However, the mean annual rainfall of Jhansi and Tickemgarh are 850 and 750 mm, respectively. Based on the collected rainfall data, both the districts has experienced declining rainfall trend (Fig. 9.1).

9.1.2 Topography and Soil

The undulating topography of this area is characterized by low hills, plateaus and plains. It is facing environment degradation with increasing soil erosion and widespread deforestation. Central Ground Water Board described this area as gradually sloping in the north-easterly direction. The southern Bundelkhand plateau area in general resumes the height ranging from about 200 m above mean sea level towards north to about 345 m above mean sea level on the south. viz (a) southern Bundelkhand pediplain province and (b) northern highly eroding composite plain province. Soil found in the area may be classified into two group on the basis of colour and topography i.e. red (upland soils) and black (low land soils). On the basis of texture, the red soil is further divided into 'Rakar' and 'Parwa' and the black soil group into 'Kabar' and 'Mar'. Soil depth in Kumharra and Azadpura varies between 2.5 to 3 m. Traditionally, groundwater resources are meagre due to the presence of hard rock like gneiss and granite. Availability of water varies from season to season. In this region agriculture is highly dependent on rainfall and only 20 % of the total area under irrigated agricultural.

9.1.3 Biodiversity

Tree and shrubs commonly found in selected area are Dhak (*Butea monosperma*), Amaltas (Cassia fistula), Anjan (*Hardwickia binata*) and shisham (*Dalbergia sissoo*), kala siris (*Albizia amara*), subabul (*Leucaena leucocephala*), babool (*Acacia nilotica*) nutan (*Dichrostachys cinerea*), Aonla (*Emblica officinalis*), Neem (*Azadirachta indica*), peepal (*Ficus religiosa*) bargad (*Ficus bengalensis*), Arjuna (*Terminalia Arjuna*), *Lantana*, *Chrysopogon fulvus*, *Zizyphus nummularia* etc.

Forests are endowed with rivers Jamni and the Betwa. These provide nature with serene environment with splendid forest and vegetation. These are habitat of *Sloth Bear, Nil Gai, Wild Boar, Langur, Spotted Deer, Jackal, and Peacock.* In these forests common vegetation are Jamun, Teak, Mahua and Achar.

9.1.4 Common Crops

Jwar, wheat, Maize, bajra, kodo, kutki, gram, tuar, mung, urad, and soyabean, alsi, til, mungfali, jute are the commonly harvested crops. *Soyabean* is also a major crop in the area which has attracted farmers due to its cash value in the market. But increasing area under Soybean cultivation has affecting the crop diversity as the area under other oilseed and pulse crops has been diverted to area under soybean.

9.2 Inferences from Primary Survey

Based on the information collected through the focus group discussions, and primary and secondary data collected, the inference drawn is discussed as follows (Figs. 9.2, 9.3, 9.4 and 9.5).

Fig. 9.2 Discussion with villagers in Ghisoli, Jhansi

Fig. 9.3 Discussion with villagers in Kumharra village, Tikamgarh

Fig. 9.4 Breached check dam in Ghisoli, Jhansi

Fig. 9.5 Step well, traditional water harvesting structure in Kumharra, Tikamgarh

9.2.1 Water Conservation

Ghisoli and Raksa villages have water harvesting structures. In Ghisoli, a small *nallah 'Mauri'* flows during rainy season. A large pond is also available in Ghisoli, at the time of visit it was having little water (Fig. 9.6). In Ghisoli various check dams are also available along *nallah*. The scemes under MNREGA has actively participated for water conservation. This has not only created infrastructure but also helped in getting employment in village. Check dam constructed by Srijan NGO. Srijan also helped for *medbandi* for avoiding soil erosion and conserving water in agricultural fields. Srijan NGO has worked here in 1995 and did the plantation work. Rajiv Gandhi Foundation is also active in this area. It helped in forming *Self Help Groups* and about 25 such small groups are functional here. Money collected is used for buying cow & buffalo. Due to high rearing cost and unavailability of pasture land, villagers do not appreciate this scheme.

Raksa has various water harvesting structures (Fig. 9.7). It has *Khaprar* dam of Uttar Pradesh Irrigation Department, *Putania bandh* and few small water storages (*talai*). In Raksa villagers informed that since last 5–6 years Khaprar dam (Irrigation Department) is not filled, otherwise irrigation problem does not occur. Khaprar dam is near Janakpur Majra of Raksa. This majra has 25 households, and here hand pumps are also working. Putania bandha (near Baghpura majra) is not filled with water since last 4–5 years. Encroachment was also noted during field visit. Villagers informed if water is filled in this dam it stays up to 6 months. A diesel pump was working for extracting water from dam bed. *Ayoya bandha*, *Bujaria talab*, check dam at *Dharan ki har majra*, and 4 other check dams are in the village. *Gumanpura majra* has also small check dam.

River Satar flows through Azadpura village and Kushanagar (Nakta). It receives overflow of Rajghat dam and a small *nallah*, river Babedi which receives water from Jhansi. These are rain fed channels. There is series of about 10 small check dams on river Satar. Azadpura and Kumharra have old *talab* and these require maintenance work (Figs. 9.8 and 9.9). In these villages no other intervention for rainwater harvesting has been done.

Fig. 9.6 Major WHS, plantation and habitation in Ghisoli

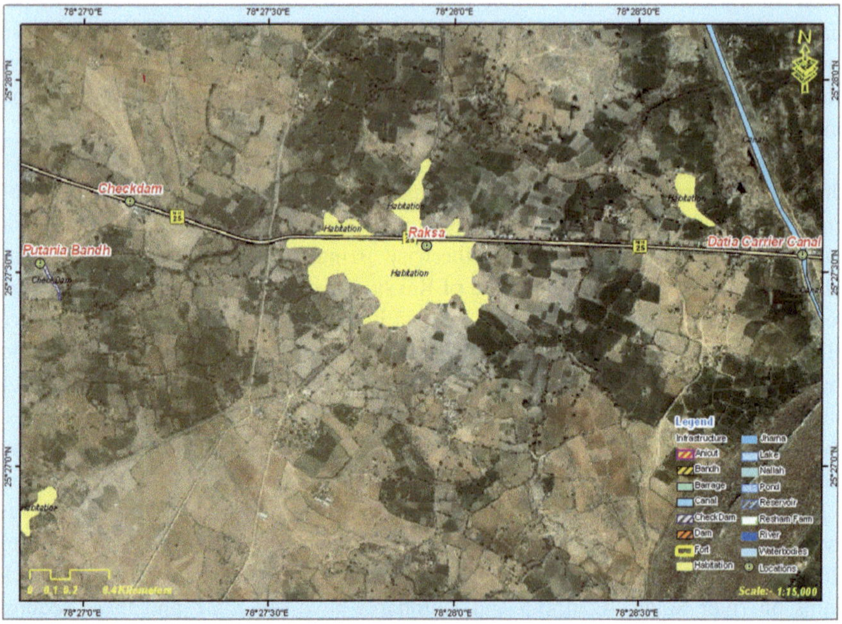

Fig. 9.7 Major WHS and habitation in Raksa

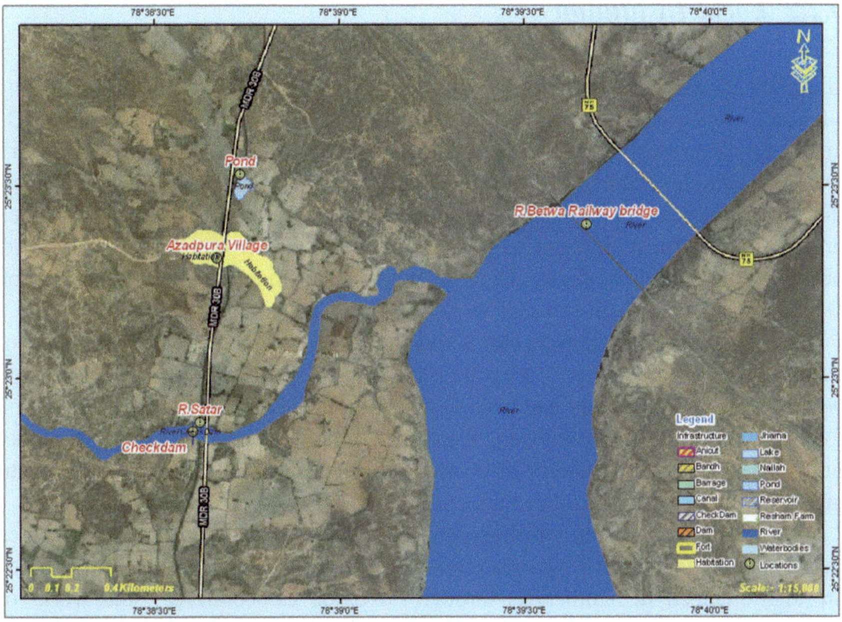

Fig. 9.8 Water bodies and habitation in Azadpura

Fig. 9.9 Major WHS, habitation and stream in Kumharra

9.2.2 Local Institution

Ghisoli had one village level institution for maintenance of water harvesting structure. However, villagers informed that it was not functional and nobody knows anything for that. However, villagers themselves have informal institutions which discuss the plans for village development. In the discussion, it came out that they have plan for maintaining the breached check dams. However, they need proper training for maintenance. Raksa village has also a water user association, but the interaction is poor. They informed that many years ago Irrigation Department helped them through workshop for crop management.

On the other side, Azadpura and Kumharra villagers are aware of importance of water conservation. Villagers are interested in the watershed development through participatory approach and had tried to discuss their problems with administration for construction a small dam/weir. Kumharra village does not have any formal or informal institution. Villagers informed that they need training about the techniques for the conservation and watershed management.

Women Cooperatives: Development Alternatives (DA) is working in this area for livelihood generations, women empowerment and water conservation. Women are also engaged in handicrafts through "Women Cooperatives" and work for weaving jute, paper products etc. Housing development for tribals (Sahariya community) is also remarkable development by this institution. These houses have been made using local material through non–conventional methods. Such initiative need to be replicate further in these villages.

9.2.3 Environmental Status

Drinking water adequacy and quality: Drinking water requirement is fulfilled by open well and hand pumps. There are about 500 well. New 8 wells have been constructed by government scheme for SC people living unde Below Poverty Line (BPL) category. Drinking water supply under Accelerated rural water supply scheme (ARWS) has provided 30 connections to Ghisoli. In Raksa also, piped water supply was facilitated by government drinking water that was sufficient for about 1.5 years in 2002–03. Raksa has 60000 liters over head service reservoir (OHSR) capacity. However, because of lowering of water table the water extraction was not possible further. So, it became non functional. Villagers informed that during construction of National Highway the pipelines were broken. After construction of the National Highway (Jhansi to Shivpuri) again people are tried for piped supply, but the earlier source is abandoned. Household living near the National Highway have troublesome life, although while visiting the area it seems better developed as compared to other villages, but drinking water problem is severe. No hand pump and well is working. People depend only on tanker water supply which comes one time a day, for which villagers are spending money.

Initially, Jal Nigam has supplied water through the tankers free of cost. Now the people purchase the water from local bodies at the cost of Rs 1 for 15 L.

Water quality is satisfactory in Azadpura for drinking and irrigation. In Kumharra, only 3 hand pump are used for drinking water requirement. Two hand pumps have saline water problem. Among the abandoned 5 hand pump, 2 hand pumps had salinity problem. Boring is successful at 23.0 m and water is available up to 10–12 m. After 24 m water is saline. Water level in Kumharra is at about 9.0 m. Water quality up to this level is sweet; however, water in lower aquifer at about 24 m is saline. Problem of water quality actually started from last 5–6 years. Drinking water availability may be enhanced through increasing the water level by rejuvenating traditional system and innovative water harvesting structures.

Drinking water requirement is usually fulfilled by the hand pumps. Saline water problem exists in this area. Few hands pump India Mark-II and boring at 50 ft. has saline water. Drinking water supply from Nagar Panchayat is received for 1 h daily through tube well which is not sufficient. However, water is in good taste and quality seems to be adequate. Azadpura has 5 hand pump out of these 4 are working. Kumharra has 10 hand pump and 5 are working. There is one private bore well which is used for drinking water.

Water Table: Raksa has 14 well used for drinking water. In Raksa water is saline and wells are dry in the habitation near to National Highway. They are suffering acute scarcity of water since last 6–7 years. Villagers informed that before 20–25 years water was adequately available for all. Water was available at 7.5–9.0 m in open wells. Wells in agricultural fields have water at 21.0 m. A few old wells also have the saline water at 15.0 m. These were saline earlier also. Table 9.1 shows declining in water level in the villages. Depth of drinking water tube wells of Central ground Water Board varies from 100 to 150 mbgl for Babina. The yield of tube wells varies from 200 to 600 lpm in hard rock areas.

Ghisoli has about 31 hand pumps, which are at level of 30.0 m. Saline water was reported at 24 m in one hand pump. Drinking water facility is available within 70–80 m radius. There are 9 well in Ghisoli which are used for drinking. Raksa has about 110 hand pumps, however, only hand pumps installed at Janakpur *Majra* is working as it is near agricultural fields. Ghisoli village as has 127 *pacca* well which are used for irrigation. They have 123 diesel engines for pumping water for irrigation.

In Azadpura, there are total 150 *kutcha* wells exists, which were used for drinking and irrigation. Out of these about 80 wells are abandoned or dry. Working open well has sweet water (good quality) at about 6.0–7.5 m. Open well at "*Amar Shaheed Chandra Shekhar Azad Agyatwas Sthal, Satar*" has water level at only 2.2 m with good quality. Another open well at this place (3.0 × 3.0 m area), is used for farming/gardening and about 1.5 m water can be drawn in one operation of about 6 h. After 24 h it again reached to its initial level. This open well site very close to the Satar River which gets water from Rajghat dam, and can be said as the river bank site. At present, Azadpura has 30 *pacca* well (only 3 are working). They have 30 electric engine pumps. On an average water table is declining in the area (Table 9.1).

Table 9.1 Water level in open well (m bgl)

District/Village	Year 1990		Year 2000		Year 2010
	Pre monsoon	Post monsoon	Pre monsoon	Post monsoon	Pre monsoon
Jhansi					
Ghisoli	13	11	15	13	17
Raksa	17	15	20	19	25
Tickemgarh					
Azadpura	12	10	16	13	21
Kumharra	15	14	18	14	24

Source Primary survey

Sanitation and hygiene: Various houses in main village Raksa and Ghisoli have in-house toilet facility. However, in majority of houses open defecation is practiced. No agency has worked in these villages for sanitation and hygiene works. However, in Azadpura and Kumharra villages has open defication system, though 2–3 families have toilet facility.

Though Raksa and Ghisoli has toilet facility but hygiene was absent because of inadequate water supply. But in Azadpura and Kumharra village has hygeinic system because nature is taking care of the hygeine.

Waste Disposal: In solid waste disposal and maintaining the hygeine in the area population is playing a bigger role. In densely populated area like Raksa, waste disposal is problematic. However, its *majra* which are located near fields do not have this problem, Raksa village need community awareness for waste disposal and better hygiene practices. Ghisoli village has comparatively better status because of awareness and open area.

In Azadpura and Kumharra village, open dumping in low lying land has been in practice. Waste disposal problem was not observed as they have barricaded the site through bushes etc.

Disease: There is no pathogenic or non-pathogenic disease is prevalent.

Fuel: In both the villages fuel wood and cow dung are mainly used. However, many household use LPG stoves also.

9.2.4 Socio-Economic Status

Population and Landholding: Ghisoli has population of about 3000 with 600 households. About 25 household have 10–12 acre of land and come under medium category of operational land holding. However, 50 % household have less than 5 acre of land. 10 % population is landless. *Thakur, Brahmin, Gadria* and *SC* people live in the village. Raksa village has population of about 11500. *Brahmin, Thakur, Gujar, Aharwal, Gadaria* and *SC* people live in the village. It has about 2000 households. 50 household belongs to landless category and about 50 household have only uncultivable land (banjar or gravely). Raksa is surrounded by undulating topography with small hillocks. Households are scattered near the agricultural lands and are called *majra.*[1] There are 12 majra in Raksa. Earlier there were a lot of families which were land less but in year 1970 village land were donated to landless villagers, which improved the status.

Azadpura has about 2100 population. Village is residence of *Rajputs, Pal, Thakur, Yadava, Lodi, Napat, Vanshkar* and *Tribals.* Village has about 500 households, out of these 60 households come under landless category of operational landholding. Tribal families mainly come under this head. There are about 30–30 household with large landholding of 40 *bigha*[2] and 25 *bigha.* Kumharra village has population of about 2700 with 520 households. About 20 households come under small category of operational landholding and have 12–13 acre land. However, majority of farmers come under marginal category of land holding and have 1–5 acre of land.

Education: Girls and boys go to school in the villages. Literacy in these villages varies between 30 and 45 %. More or less, these village have the accessibility of secondary education.

Occupation: Most of the villagers are working as farmers, agricultural labourers and pebble-gravel mining work. Very few are working with government organization in near by the area like Jhansi and Tickemgarh. People also go to Delhi, Agra and Jhansi for working as labourer. Skilled villagers works as artisans in cities. In Azadpura, some ladies work for handicrafts at Taragram (DA). In Kumharra village, few landless villagers are shepherd and migrate to nearby areas with cattle and come back after rains.

Migration: Because of insecure agricultural income, villagers usually migrate to Jhansi, Agra and Delhi. Since agriculture income is not secure, therefore, villagers used to go nearby cities for employment.

Status of Women: Women usually spend time in drinking water and fuel wood collection. They also support in agricultural activities. In few villages, women have been helped by non-government organization for cooperative development and skill development to work for handicraft. However, only 5 % women are aware of development works.

9.2.5 Agriculture and Irrigation

Agriculture: *Moongfali, moong and makka* in kharif and *wheat, chana* and *sarson* in *rabi* are main crop of Ghisoli (Figs. 9.10 and 9.11). In the Raksa village, similar cropping pattern exists; however, *urad* and *moongfali* dominate over the other crops during *kharif* and wheat during *rabi* (Figs. 9.12 and 9.13).

---

[1] Small habitation of a village.

[2] 1 Acre = 1.75 bigha.

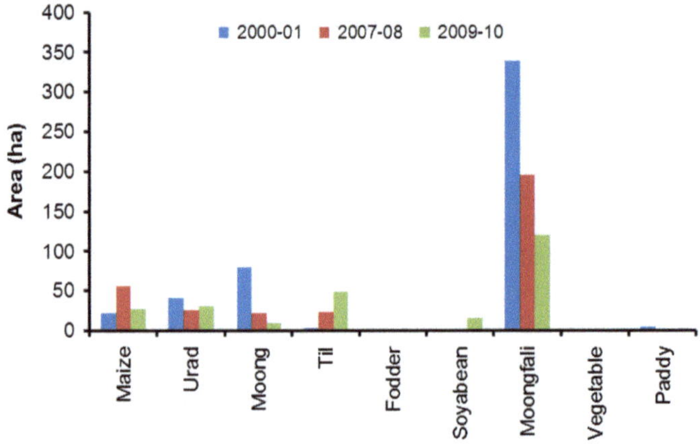

Fig. 9.10 Cropped area during Kharif in Ghisoli

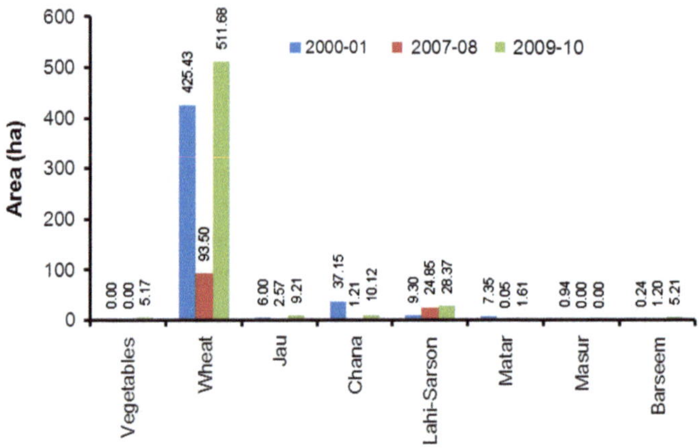

Fig. 9.11 Cropped area during Rabi in Ghisoli

Crop Production: Villagers informed about the crop production that *wheat, makka, chana* and *sarson* are main crops for their survival. Ghisoli has better crop production than Raksa, this may be because of capacity building through institutional development at Ghisoli and farmers are more aware for better crop management practices. Villagers informed that ovel all crop production has improved (Table 9.2). However, they relate it to the usage of high yielding variety and fertilizer application.

Azadpura has total area of 287.99 ha out of this 35 % is cultivable area and 39.6 % forest area (Fig. 9.14). In Rabi season wheat is the main crop. During Kharif, *ground nut, til, urad, arhar* and *chana* are grown.

Kumharra has 405.17 ha total area out of this 65 % is cultivable area. In Kumharra, villagers take two crops in a year. In Rabi, *wheat, chana, sarson, matar,*

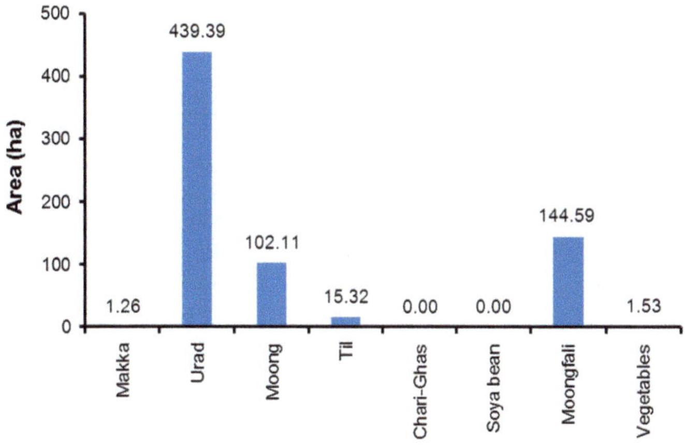

Fig. 9.12 Cropped area during Kharif in Raksa (2003–04)

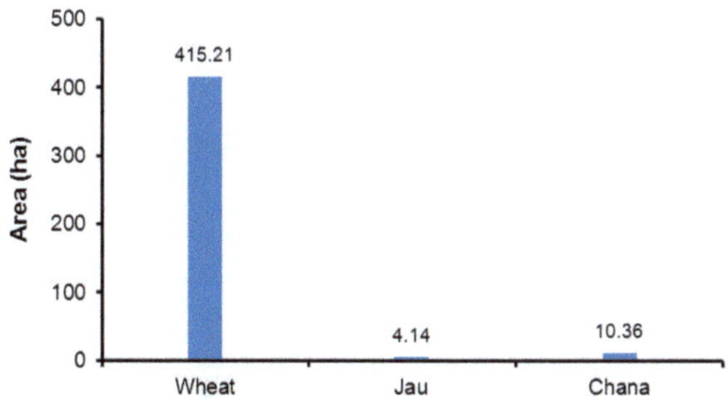

Fig. 9.13 Cropped area during Rabi in Raksa (2003–2004)

and in Kharif, *moongfali, til, urad, moong, makka*, and *soybean* are grown. Among vegetables, *bottle guard, ladies finger, tomato, brinjal, carrot, reddish* is cultivated. Figures 9.15 and 9.16 show cropping area for *kharif* and *rabi* in these villages.

Villagers buy seeds from the market, very few families store seed. Situation during low rainfall year is very alarming. Villagers informed that during drought year before two years only 5 % of agricultural area was cultivated.

Irrigation: In Ghisoli irrigation is done by well. Sprinkler irrigation is usually beneficial for such areas, but is not successful due to faster drawdown rate within 4–5 h. Irrigation is mainly done by tube well and open wells. Canal irrigation is very less. Villagers informed that they are against the order of unauthorized water supply to the farmers in nearby area from *Patraul* Irrigation Office. About 487.3 ha area is irrigated in Ghisoli. Canal irrigation is done in 12.5 and 27.3 ha through pond, rest is irrigated by well and tube well. Raksa has 97 well which are used for irrigation. They have 110 diesel engines. Village also has 118 *kutcha* well, few of which are used for irrigation. Canal length is Raksa is about 2 km.

Table 9.2 Crop production in the selected villages (kg/ha)

Crops	Year 2000		Year 2005		Year 2008	
	Ghisoli	Raksa	Ghisoli	Raksa	Ghisoli	Raksa
Wheat	2500	2400	2800	2500	3000	2800
Makka	550	540	600	600	600	600
Urad	550	600	550	550	600	600
Moong	550	600	550	550	600	600
Jau	50	70	70	70	65	70
Chana	550	600	550	600	600	600
Sarson	750	750	800	750	800	800
Moongfali	540	550	700	800	900	900

Source Primary survey

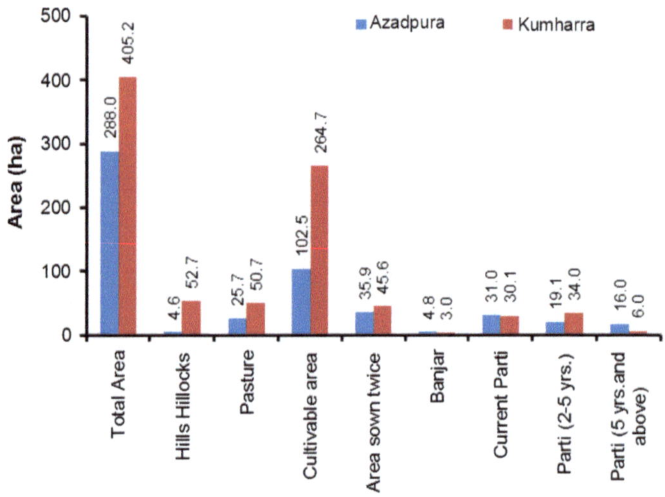

Fig. 9.14 Land use in Azadpura and Kumharra (2008–2009)

Fig. 9.15 Cropped area during Kharif in Azadpura and Kumharra (2008–2009)

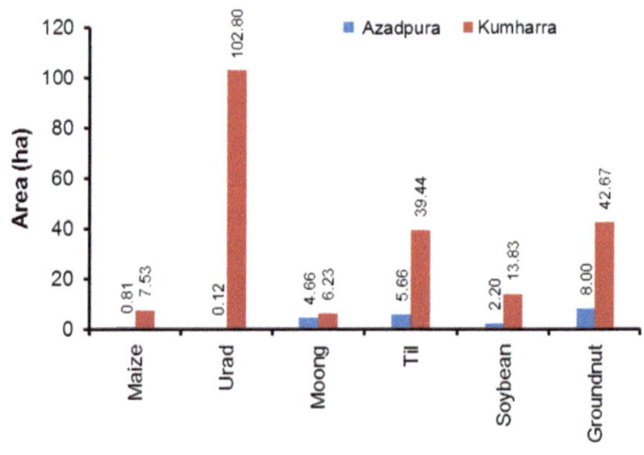

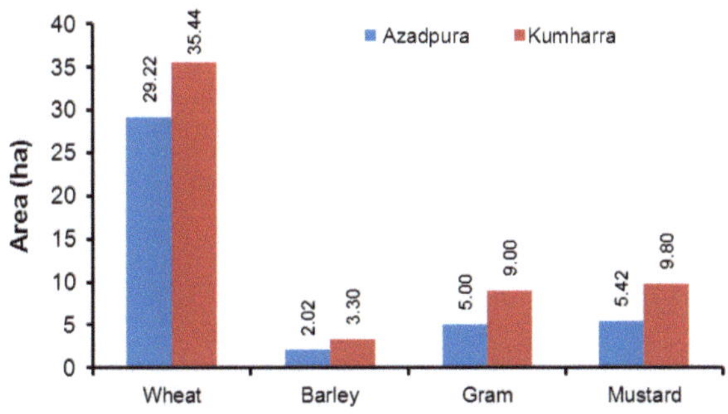

Fig. 9.16 Cropped area during Rabi in Azadpura and Kumharra (2008–2009)

In Azadpura and Kumharra, irrigation is done by well and tube well. Earlier they used to irrigate by pond water also. The pond is now broken and does not allow water to store. Water quality for irrigation is agreed.

Livestock: Every household have livestocks. Cattle rearing are mainly for in house milk consumption. In Raksa villagers informed that they also sell milk in local market. Cattle rearing were earlier high due to less cultivated area. However, dependency on agriculture has now increased and cattle rearing are considered as costly affair. This is mainly due to lack of pasture. Villagers informed that about 20 years back they had more animals and insecurity for food was not there. But due to consequent drought years they sold cattles, and few families left them open as they had nothing to feed them. Villagers informed that they do not have land to develop pasture and forest department has restricted entry of cattle. Presently, livestock population is very low and has reduced from year 1990 to 2010 (Table 9.3).

Farm Power and Machinery: People use tractor for ploughing in Ghisoli, very few families use bullock. Farmers have own agricultural equipments such as wood plough, steel plough, cultivator, leveller, thrasher and winnower.

Fertilizer and pesticide application: Chemical fertilizer and organic manure are used as manure. However, presently, farmers are more dependent on the chemical fertilizes as livestock population is getting down in recent years; and therefore, the application of pesticides and fertilizers has been increased.

9.2.6 Amenities

School and Hospital: There is primary and middle education facility in these villages, though secondary education facility exists in few villages, which are near to the cities. Apart from the school, other infrastructures available are *Kisan Seva Kendra, Gram Panchayat Karyalaya, Swasthya Kendra* and *Sadhan Sahkari Samiti*. In spite of having large infrastructure available in village, literacy is low.

Road: Villages are well connected to *pacca* road and National Highway.

Table 9.3 Livestock population in Ghisoli and Raksa

Livestock	Year 1990	Year 2010
ghisoli		
Cow	150	100
Buffalo	170	150
Goat & Sheep	2000	1500
Raksa		
Cow	180	150
Buffalo	200	180
Goat & Sheep	3000	2000
Azadpura		
Cow	150	100
Buffalo	200	170
Bullock	200	50
Goat & Sheep	10000	3000
Kumharra		
Cow	150	80
Buffalo	200	150
Bullock	200	30
Goat & Sheep	6000	3500

Source Primary survey

Electricity: Villages has electricity supply. About 8 to 10 h daily power supply is available.

9.3 Institution

The schemes under MNREGA have worked in Ghisoli area for pond development through public participation. Forest Department is also active in the area and has done plantation and soil conservation work. Department of Resham, Government of Uttar Pradesh has developed nursery in Ghisoli for *Arjuna* tree for sericulture. Tusser silk is produced from this.

Rural Drinking Water and Sanitation (SWAJAL) Project was implemented here under the World Bank Scheme, located in Jhansi and Lalitpur districts covered 26 villages. The project was conceived in consultation with the project beneficiaries and was implemented through the formation of users group in different villagers headed by women members. The main components of the project included—non formal education, training on hygiene, environment and sanitation awareness, women's development initiatives. Similar scheme is needed in these villages especially in Raksa where drinking water problem is severe.

Forest department of Madhya Pradesh has also worked for watershed management in forest area; however, that does not have any remarkable impact in this area. Forest department may intervene in village land also so that water conservation can be done on watershed management approach. A non-government organization "Sarthak—Friends of Orchha" is demonstrating eco-friendly techniques of construction and sanitation in its home-stay project and involving youth in campaigning for the protection of the environment hardly 5 km away from this site. In constructing the home-stay rooms, particular attention has been paid to ventilation and insulation. The brick walls have been constructed using the "rat-trap" technique that leaves hollow spaces within the wall that help keep the room cool in summer and warm in winter. This organization has also helped in constructing bathroom and toilet that allows water and waste to be recycled. Waste water from the baths passes through a sand filter and is then used for irrigation. Four of the houses have dry, composting toilets built according to a German design technically called UDD (Urine Diversion and Dehydration). These produce excellent compost that can completely replace chemical fertilizers. In families that possess animals, the toilet is linked to a bio-gas digester. Excreta and cow dung are decomposed to produce methane that is used for cooking as well as organic fertilizer.

9.3.1 Capacity Building

Intervention by government and non-government organization has helped the villagers in developing awareness and capacity building for natural resource management. Srijan worked here for watershed development and its work is still effective. Villagers informed need of such more works in Ghisoli so that erosion during monsoon can be checked and water can be conserved. There are many broken structures which need maintenance. It has been observed in the village that after the project completion, local institutions developed for operation and maintenance of structure has been defunct. Therefore, need of awareness and training for IWRM approach is required before such intervention works so that longevity of infrastructure created increases.

Farmers need better farming techniques for ploughing, irrigation and fertilizer application. In the area, as such no organization has worked for the awareness and training to farmers for better cropping and watershed management aspect.

9.4 Summary and Conclusions

The summarized recommendations arrived at from the study are presented in Table 9.4. The survey/campaign in Sind and Betwa catchment shows poor institutional role in the holistic socio-economic development and natural resources management. Government has not taken up any initiatives for sustainable

Table 9.4 Summarized status and possible recommendation of the issues

S. No.	Item	Status	Concluding Remarks/Recommendation
1	Education	School and college available	Vocational training centers are required for sustainable socio-economic growth
2	Drinking water	Drinking water scarcity	Drinking water facility needs improvement
		Salinity problem	
		Ground water level is lowering	This region is rich of river systems, for groundwater augmentation river bank should be used
			Topography supports for the surface water storage. Similar to Banganga catchment, traditional water harvesting system need to be encouraged
3	Sanitation and Hygiene	Open defecation is practiced by majority of population. Some families have in-house toilet facility in Ghisoli and Raksa	Community toilet may be proposed. Awareness and training need to be provided so that these are maintained. These need water supply for maintaining hygiene
			In less densely populated area, open defecation system is not creating any problem to the hygiene
4	Electricity	Frequent power cuts	Regular supply should be maintained so that they can manage the water requirement through tube well. Alternate energy should be promoted like solar energy for household and irrigation with subsidized rate. It will definitely encourage the application of efficient irrigation system and therefore, the irrigation losses will be reduced
5	Land resources	Severe soil erosion on slopes and along *nallah*	Check dam are required at many location and existing one needs maintenance
			Forest Department should develop a catchment area treatment plan and according to the erosion vulnerability, either biological or structural measures of soil erosion control should be adopted. This will also control the reservoir sedimentation in the catchment or river
6	Forest	Deforestation	Forest should be protected and wherever required plantation should be done. It will protect the soil loss from the catchment
		Afforestation activity is going on	

(continued)

Table 9.4 (continued)

S. No.	Item	Status	Concluding Remarks/Recommendation
7	Water Resources	High surface runoff	Runoff is very high during rains. This leads to soil erosion and loss of water from the catchment
		Maintenance is required for existing harvesting structures	Additional water harvesting structures and soil conservation structures and maintenance of existing ones is required
			Construction of small tanks and reservoirs at suitable sites particularly in run off zones
			Construction of small check dams and bunds in major and minor nallah to check flow and to recharge the ground water
8	Agriculture	Low crop yields	Soil conservation measure and usage of organic manure should be increased
		Increasing cultivation cost	
		Low crop diversity	Crops with low water demand should be promoted. This will reduce the cost of cultivation and over exploitation than recharge
			Crop diversity should be enhanced for secure economy of village as well as better crop- water management
			Department of Agriculture should develop a cropping plan at micro-watershed level considering the economy, water availability and market facility for sustainable growth
9	Irrigation	Well	Water level in lowering
		Tube well	
		Canal	Inadequate canal water supply
			Water harvesting structures were observed for pumping water
			Water quality is all right, however extraction of water is very high

(continued)

Table 9.4 (continued)

S. No.	Item	Status	Concluding Remarks/Recommendation
10	Livestock	Fodder management issues not addressed	Livestock are major part of rural economy. Integrated water resource management addresses the whole requirement of a watershed. However, in spite of having water harvesting structure impact is not very high, because of no holistic approach
			Plantation and regeneration of fodder tree needs to be prioritized
			Protected pastures and farm forestry land may be developed
			For IWRM, mixed farming needs to be encouraged. There should have balance between livestock and crop production
11	Non Farm sector	Not scaled up	Non-farm activities may include handicraft, embroidery etc. Women may be encouraged to work as these may be put up for sale by some small agency in nearby places which have high tourism value (Jhansi, Orchha, Khajuraho)
12	Economic security	No stabilized income Risk during aberrant weather	Crop diversity, pasture, farm forestry and non- farm activities should be implemented for stabilize income
13	Institutions	Water user association Self help groups	Water user association in the both the village has no significance at present. Because of limited scope and no proper training and awareness there never worked
			Villagers were trained for water conservation and better cropping by Irrigation Department in Raksa and Srijan in Ghisoli
			They may be revived and trained for better livestock and farming & non-farm activities for revenue generation
14	Landscape	Landscape transformation	Deforestation and frequent drought years has damaged the rejuvenation. Forest department is actively working for plantation and forestation. Local institutions effectively can work for sustaining forests. Awareness about the changing effects of land use change, pressure on agriculture and irrigation can serve as a useful tool in formulating regionally appropriate policies

watershed management though public is keen to participate in the development programmes. Government has done some works under MNREGA but mostly the pond development without technical understanding of the system. Soil erosion is also a major concern in catchment, but government has not taken-up any initiatives. Considering these facts, there should have IWRM plan for the catchment at micro-watershed level.

References

Jain SK, Agarwal PK, Singh VP (2007) Hydrology and water resources of India. Springer, New York

Rai RK, Upadhyay A, Ojha CSP, Singh VP (2011) The Yamuna river basin: water resources and environment. Water Science and Technology, vol 66. Springer, The Netherlands

Chapter 10
Case Study: Ken Catchment

Abstract In spite of various perennial rivers flowing in the catchments, water scarcity is persisting due to lack of integrated land and water management practices. Government has initiated some Programmes but failed because of non-technical approach and absence of clear-cut vision. The water conservation work to satisfy certain class/caste was noticed, which violates the IWRM principles and public cooperation and participations.

The Ken River is one the major rivers of the Bundelkhand region of central India, and flows through two states, Madhya Pradesh and Uttar Pradesh (Jain et al. 2007). The Ken River originates near village Ahirgawan on the north-west slopes of Kaimur Range in Jabalpur district and travels a distance of 427 km, before merging with the Yamuna at Chilla village, near Fatehpur in Uttar Pradesh. Ken has an overall drainage basin of 28,058 km^2, out of which 12,620 km^2 belong to Sonar River its largest tributary, whose entire basin lies in Madhya Pradesh; and along its 427 km course it receives water from its own tributaries such as Bawas, Dewar, Kaith and Baink on the left bank, and Kopra and Bearma of the right. The upper reaches of Ken River are flanked by undulating plateau with sandstone, shale and limestone. Down below recent alluvium engrosses the river up to Daudhan dam site. The stratigraphy of rock formations found in the region is mostly alluvial soil, Deccan traps, Lameta beds and Vindhyan system. Crossing the Bijawar-Panna hills, the Ken River cuts a 60 km long, and 150–180 m deep gorge. Several streams join the Ken in this gorge making waterfalls. The Ken valley separates the Rewa Plateau from the Satna Plateau. The basin covers the areas of Jabalpur, Sagar, Damoh, Panna, Satna, Chhatarpur and Raisen districts of Madhya Pradesh and Hamirpur and Banda districts of Uttar Pradesh (Rai et al. 2011).

In the study, Banda district was selected. The ground water of Banda is declining at the rate of 65 cm every year, the fastest rate in northern India.

A. Upadhyay and R. K. Rai, *Water Management and Public Participation*, SpringerBriefs in Earth Sciences, DOI: 10.1007/978-94-007-5709-7_10,

Fig. 10.1 Discussion with villagers in Ragauli Bhatpura, Banda

According to CGWB, water of Banda district is saline at various locations (CGWB 2010). Two villages of its Naraini block namely the Ragauli Bhatpura (Fig. 10.1) and the Masauni Bharatpur was identified to investigate the water management and public participation (Figs. 10.2 and 10.3). The Ragauli Bhatpura village has severe drinking water and irrigation problem; whereas the Masauni Bharatpur has natural gift for the excellent scope of water harvesting as this area is flat surrounded by hills. Both the villages have two small rain harvesting structure. Masauni has very old wells of the time of kings, a 127 years old is still functional at the time of visit with good quality of water.

These areas are drained through the river Baghein, Paisuni and Gadra nala, which are the tributaries of Yamuna. This river had a general tendency of river bank erosion and scouring, especially the southern bank, which causes displacement and destruction of many villages. A famous village Shaipur near Chilla-ghat the headquarters of Pargana Pailani during Moghul time is said to have been entirely swallowed by it (www.banda.nic.in). Flowing circuitously towards north, south and south-east directions Yamuna is joined by Ken at Chillaghat, Bhaghein near Bilas, and Paisuni near Kankota villages.

River Baghein arising from a hill near Kohari of Panna district enters Banda district at Masauni Bharatpura village. It flows north-east-ward and at a point separates Banda from the newly created Chitrakut district forming boundaries between Atarra, Baberu and Karwi tahsils. It being most capricious in its action, deposits quantities of sand, but near its junction with Yamuna it tends to flood a large area of low lying land, if the stream in the Yamuna is sufficient to block its outlet. The chief tributary of Baghein, river Ranj, joins it at Gurha Kalan of tahsil Naraini but further east; there are several smaller tributaries from south namely river Madrar, Barar, Karehi, Banganga and the Barua, each of which in turn has tributaries of its own. River Barua has been dammed to provide some irrigation through canals.

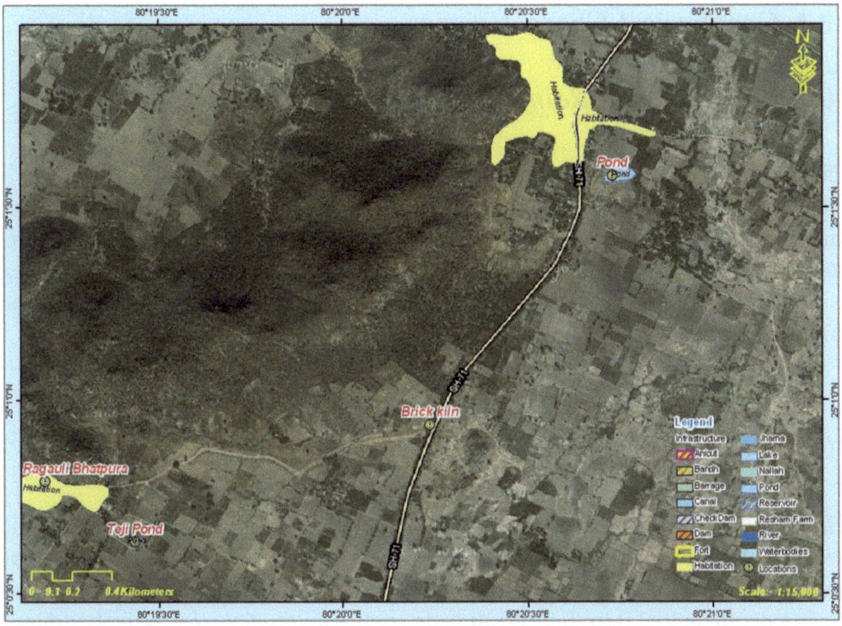

Fig. 10.2 Major WHS and habitation in Ragauli Bhatpura

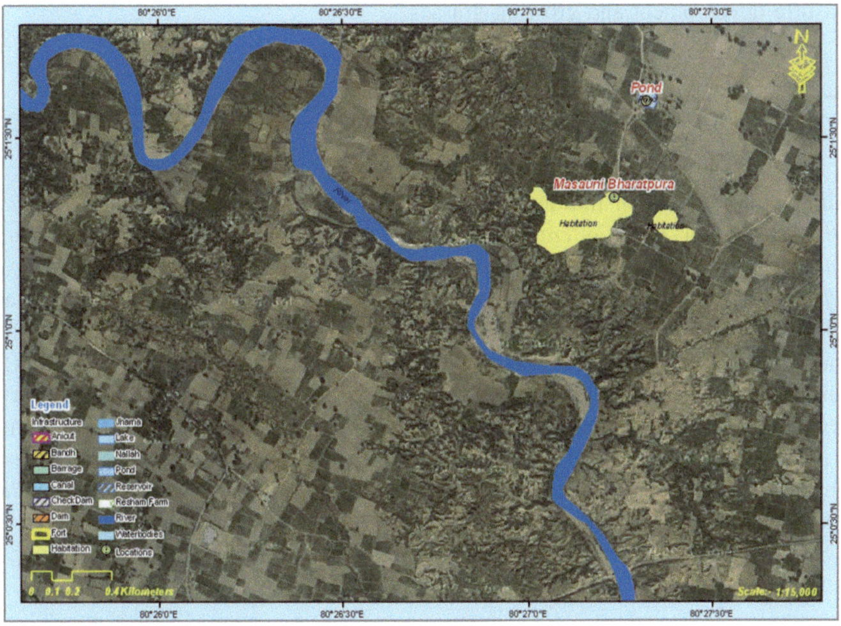

Fig. 10.3 WHS, habitation and river Baghein in Masauni Bharatpur

10.1 Geophysical Characteristics

10.1.1 Climate and Rainfall

The normal annual rainfall of the Ken catchment up to Daudhan dam site is 1,174.0 mm. The monsoon rainfall constitutes as much as 90 %. The maximum, minimum and the average annual weighted rainfall values are 1,662.1 mm, 660.0 mm and 1,181.3 mm respectively (www.nwda.gov.in). Banda has average rainfall of 900 mm, and maximum and minimum temperatures of 44.4 °C and min. 5.2 °C.

10.1.2 Topography and Soil

The area largely consists of irregular uplands with rock outcrops intermingling with mostly lowlands, frequently under water during rainy season. There are numerous rivulet/*nallah*. In Masauni soil is red-yellow. Soil is rich in nutrients having depth up to 3.0 m. In Ragauli soils are highly erodible, excessively drained and gravelly in nature. These soils are of shallow depth with low nutrient with less organic content and have poor water retention capacity. These soils are unsuitable for normal cropping. However, they may be utilized for the purposes of forestry, pasture development and growing grasses.

10.1.3 Biodiversity

Main plant species in this area are Neem (*Azadirachta indica*), Babul (*Acacia Nilotica*), khair (*Acacia catechu*), palas (*Butea monosperma*), ber (*Zizyphus varieties*), tendu (*Diospyros melanoxylon*), mahua (*Mahuca Indica*), *Embleca officinalis*, semal (*Salmalia malabarica*) and kardhai (*Anogeissus pendula*, salai (*Boswellia serrata*), seesham (*Dalbergia sissoo*), dhau (*Anogeissus latifolia*), jamun (*Syzizium*), karaundha (*Carissa spinarum*), dhawda (*Anogeissus latifolia*), tinsa (*Augenia delbergiodes*) and bija (*Pterocarpus marsupium*). However, very few trees can be seen in the village. These areas had been a dense forest along the Vindhyan range of hills. Half a century back, it was full of diverse variety of trees, bushes and herbs. Among wildlife animal species at present *deer, chinkara, wild cat, jackal, fox, wild boar* are common.

Fig. 10.4 Old well in Masauni

Fig. 10.5 WHS in Masauni

10.1.4 Common Crops

Paddy, arhar, sugarcane, maize, moong, til, groundnut and soyabean are cultivated during Kharif, and wheat, gram, masur, barley, mustard, pea and alsi are the main crops harvested in rabi season.

10.2 Inferences from Primary Survey

10.2.1 Water Conservation

Ragauli is located near the hilly area and high runoff rate from hills during rain damages the agricultural fields and therefore, water scarcity always persist in spite of good rainfall (Figs. 10.4 and 10.5). Ragauli has an old pond named as *Tezi baba* pond, which was rehabilitated under MNREGA. Village has various locations which need intervention so that water and soil conservation can be done. However, status of water conservation structure is very poor as it is not sufficient for the village.

Masauni has two ponds for water harvesting. One of them has been improved recently by Panchayat through public participation. This village has better geomorphology for agriculture. However, the village has no motivation and awareness regarding rainwater harvesting and government has also not taken up any such initiatives. In the recent work of rehabilitation, wide catchment for a *talab* has converted into a new small *talai* having very less capacity than the storage potential. Rehabilitation works were going on without any technical background causing inefficient water conservation mechanism and loss of money.

10.2.2 Local Institution

Villagers in Ragauli informed that no formal or informal institution is active in this area. However, they have shown interest in getting awareness and training for better crop management practice, rain water harvesting and soil conservation techniques. Masauni also have no such institution. However, villagers have 6–7 self-help groups for *micro financing*. Masauni is more developed because of better crop production and many of its villagers work for masonry work in Delhi, Karnataka and Andhra Pradesh. Also, village being a minority village gets benefits from Panchayat for development. Villagers informed that hand pump installed by Jal Nigam under rural drinking water supply scheme are working and water quality is also all right.

10.2.3 Environmental Status

Drinking water adequacy and quality: Villagers in Ragauli informed that the area has persistent water scarcity. By small water conservation developments they were able to retain water in their well. However, because of consequent drought years, drinking water problem has increased very acute for human beings and livestock. No pond or *nallah* is having water so that even animal can drink easily. They do not have any system or structure for providing drinking water to cattle. Water quality of open well and hand pump is satisfactory. They have three hand pumps and only one was working, which is not sufficient for whole village. They collect drinking water from nearby hand pump placed at a distance of 1.2 km. Water quality in both the working hand pump is all right.

Water table: Ragauli has 40 wells, out of which only 2–3 well are working located nearby the agricultural lands, which have hardly 1 m of water in the well. They have 70 engines for pumping water from well. Masauni has about 100 wells

Table 10.1 Water level in open well in Ragauli and Masauni (m bgl)

Village	1990		2000		2010
	Pre monsoon	Post monsoon	Pre monsoon	Post monsoon	Pre monsoon
Ragauli Bhatpura	25	23	26	25	28
Masauni Bharatpur	19	16	22	18	22

Source Primary survey

out of these 12 are working. Villagers informed that tube well are also not functional because of deep water table. Ragauli has total 12 tube well and only two are working. Masauni has two working tube wells, however, for irrigation water is insufficient. Table 10.1 shows depletion in water level in the selected villages from year 1990 to 2010.

Sanitation and hygiene: In Ragauli open defecation is practiced and no house has in-house toilet facility. Whereas, in Masauni few houses have in-house toilet facility, however, majority people practice open defecation.

Solid waste disposal: Ragauli and Masauni both the villages have open land dumping practice of waste. They have barricaded the area with bushes so that it does not spread with wind and water. Both the villages appear neat and hygienic.

Disease: Ragauli and Masauni do not witness any waterborne or communicable disease.

Fuel: In Ragauli fuel wood and cow dung are used for cooking. In Masauni 8–10 household have LPG stoves. However, majority of families depend on fuel wood and cow dung.

10.2.4 Socio-Economic Status

Population and landholding: Ragauli Bhatpura has population of about 1,500 with 160 household. About 7–8 household have 12–14 bigha of land. Other household have 1–5 bigha. About 10 percent households are landless. Masauni Bharatpur has 1,850 population and 180 household. Villagers are marginal to small farmers. About 15 household come under medium and 23 household under large landholding category.

Education: Only 20 percent population is literate. There are two schools in the village one is primary and another up to class 8th. Masauni also has a *madarsa*.

Occupation: Villagers are mainly farmers. Few work as agricultural and masonry laborer. Village elders informed that earlier there were mainly shepherd and few farmer families. Both the village have benefited with the *Swarozgar Guarantee Yojana* of Government and National Rural Employment Guarantee Scheme.

Migration: Both the villagers have very high migration. Ragauli villagers work in stone crushers in Banda as they are not skilled enough so they work in nearby areas only. Masauni villagers prefer to go Delhi, Agra, Secunderabad and Karnataka.

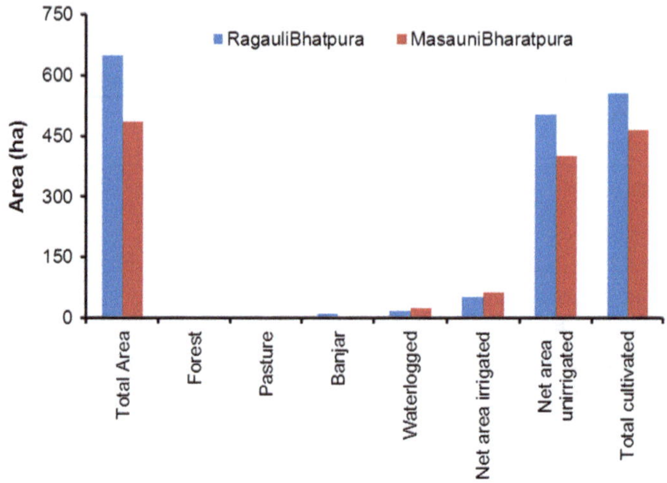

Fig. 10.6 Land use in Ragauli and Masauni (2003–2004)

Status of women: Women perform household work and agriculture in Ragauli. Women are mostly not vocal in both the villages and are illiterate. New generation of girls is reading in school. Literacy is about 15 percent for women. Women informed that they face problem for medical facility as no hospital is available nearby and they have to go Naraini for medical facility.

10.2.5 Agriculture and Irrigation

Total topographic area of Ragauli and Masauni are 647.7 and 484.42 ha, respectively. The land use pattern of both the villages is shown in Fig. 10.6. Paddy, *arhar* and sugarcane are main kharif crops in Ragauli. Cropping area under paddy has reduced in kharif and wheat in *rabi*. Whereas, cropped area under gram has been increased in Masauni. Area under *paddy* and *arhar* crop reduced from year 2001 to 2003. However, cropping area for *til, makka* and *sugarcane* increased during these years in Ragauli. Figures 10.7, 10.8, 10.9, and 10.10 shows cropping area for kharif and *rabi* for different crops. Cropping area for gram has been increased in both the villages.

Crop Production: Crop production in both the villages has improved (Table 10.2). Villagers informed that crop production has increased because of good quality seeds and application of fertilizer in both the villages. Masauni has higher production for crops as compared to Ragauli. Masauni is geographically better located for agriculture. They use good quality seeds and because of generating revenue through working as laborer they have higher purchasing power for fertilizers etc.

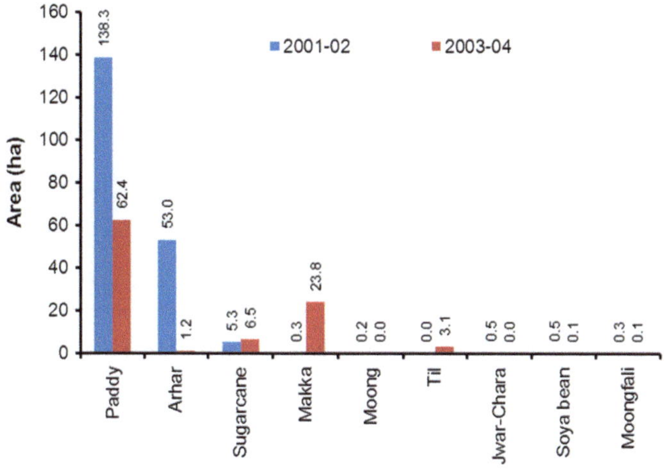

Fig. 10.7 Cropping area during Kharif in Ragauli Bhatpura

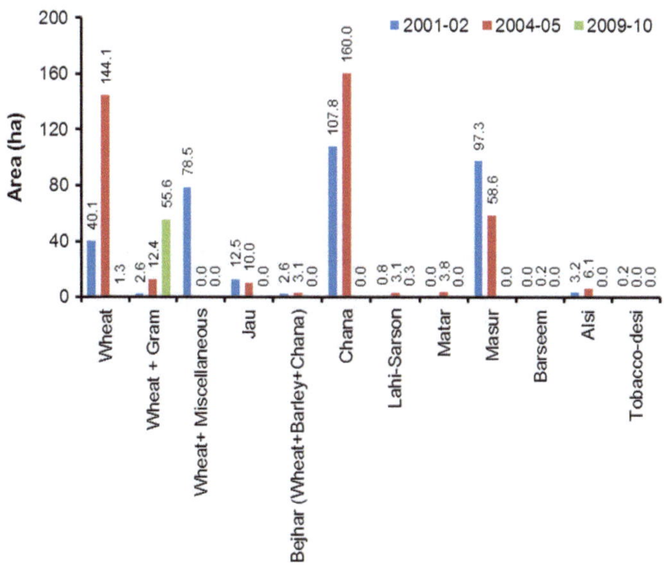

Fig. 10.8 Cropped area during Rabi in Ragauli Bhatpura

Fertilizer and Pesticide Application: Chemical fertilizers are generally used in both the villages for nutrient supplement. Villagers informed that pesticide application has increased because of usage of hybrid seed. Villagers informed that use of high yielding variety (HYV) requires more fertilizer and water; however, due to more production they prefer these.

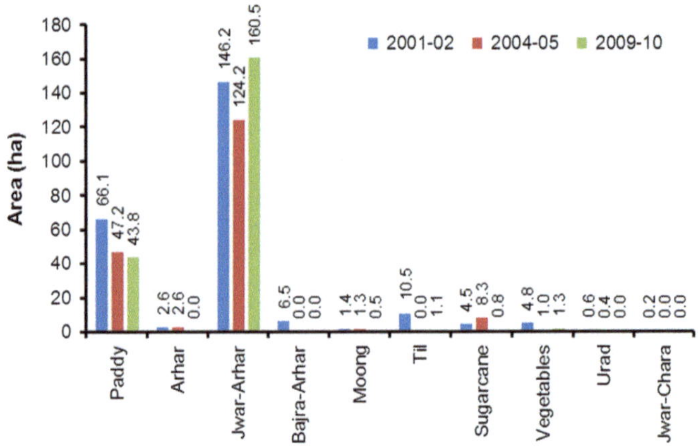

Fig. 10.9 Cropping area for Kharif in Masauni Bharatpur

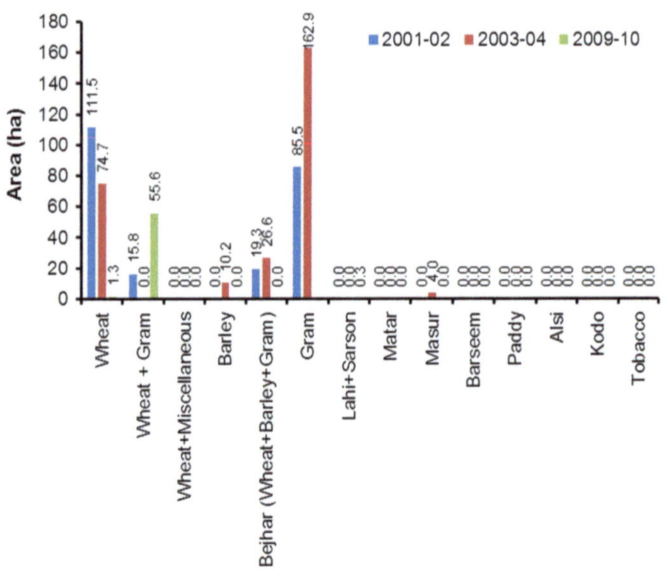

Fig. 10.10 Cropping area for Rabi in Masauni Bharatpur

Irrigation: Irrigation is done by private tube well in Ragauli. There are seven well which are used for irrigation in Masauni. They have ten diesel engines. Villagers informed that they do not have sufficient water for irrigation.

Farm power and machinery: Villagers have wood plough and steel plough in Ragauli. They have other required appliances for farming also. No household has tractor. They borrow from nearby villages. Few families have bullock however, it is negligible as per the requirement for ploughing. Masauni villagers have thrasher

Table 10.2 Crop production in the Ragauli and Masauni (kg/ha)

Crop	Year 1998		Year 2005		Year 2008	
	Ragauli	Masauni	Ragauli	Masauni	Ragauli	Masauni
Wheat	2,000	2,400	2,000	2,500	2,100	2,400
Paddy	2,500	2,900	2,400	3,000	2,500	3,000
Arhar	500	560	550	600	500	600
Maize	450	500	450	500	450	540
Jau	800	800	800	850	900	1,000
Chana	500	600	500	600	540	650
Sarson	700	860	750	850	900	1,000
Tobacco	400	400	500	500	560	600

Source Primary Survey

Table 10.3 Livestock population in selected villages

Livestock	Year 1990	Year 2010
Ragauli Bhatpura		
Cow	80	60
Buffalo	120	100
Goat and Sheep	3,000	2,000
Masauni Bharatpur		
Cow	80	100
Buffalo	100	150
Goat and Sheep	3,000	3,000

Source Primary survey

and winnower also. Tractor is mainly used for ploughing. 7–8 villagers have tractor in Masauni.

Livestock: Each household in Ragauli keeps cow or buffalo for milk. Livestock population was high before 25–30 years. Consequent drought has led to reduce the cattle rearing as adequacy of drinking water and pasture has also reduced (Table 10.3). In Masauni all household have cattle. They keep cow, buffalo, goat and sheep. Earlier they used to keep bullock also for ploughing. But now they prefer tractor. However, very few families keep bullock.

10.2.6 Amenities

School is available in both the villages. Masauni has madarsa also. Hospital facility is not available in both the villages. Mobile connection is limited in the village, and only 3–25 households have mobile in Ragauli and Masauni, respectively. Both the villages are well connected to *pacca* road. Electricity supply is available; however, 10–12 h power cut is normal.

Table 10.4 Summarized status and possible recommendation of the issues

S. No.	Issue	Status	Concluding remarks/recommendation
1	Education	Inadequate amenities	Schools as well as vocational training centers are required
2	Drinking water	Drinking water scarcity	Drinking water problem in Ragauli. Abandoned dug wells may be cleaned for usage
		Ground water (GW) level is lowering	GW utilization at sustainable basis should take place in that area Artificial GW recharge technique should be implemented through public participation. a simple technique that could be incorporated by
3	Sanitation and Hygiene	Open defecation	Community toilet may be proposed. This need proper awareness and water supply
4	Electricity	Frequent power cuts	Daily supply should be maintained
5	Land resources	Severe soil erosion in Ragauli	Conservation measure required at many locations Some curative measure needs to be taken to such as check dams and gully plugs etc Plantation is also required to check the erosion
6	Forest	Deforestation	Deforestation has aggravated the erosion and catchment destruction Villages should have forest and pasture for livestock grazing and other by products for revenue generation Protected area should be developed for limited use so that the regeneration capacity of the forests is achieved. This will also help in checking encroaching forest of Forest Department
7	Water Resources	High surface runoff Very few water harvesting structures	As the area is hilly the runoff is very high during rains. This leads to soil erosion and loss of water from the catchment Rain water harvesting structures and soil conservation structures may be proposed
8	Agriculture	Low crop yields Low crop diversity	Soil conservation measure and usage of organic manure should be increased Crops with low water demand should be promoted. This will reduce the cost of cultivation and over exploitation than recharge Crop diversity should be enhanced for secure economy of village as well as better crop- water management

(continued)

Table 10.4 (continued)

S. No.	Issue	Status	Concluding remarks/recommendation
9	Irrigation	Tube well and dug well	Insufficient irrigation water Water quality is all right, however, extraction of water is very high
10	Livestock	Pasture land reforms	Livestock are major part of rural economy. Integrated water resource management addresses the whole requirement of a watershed Plantation and regeneration of fodder tree needs to be prioritized Protected pastures and farm forestry land may be developed
11	Non Farm sector	Not scaled up	Non-farm activities need to be developed
12	Economic security	No stabilized income Risk during aberrant weather	Crop diversity, pasture, farm forestry and non- farm activities should be implemented for stabilize income As the area is at high risk of erosion, some secure income intervention is urgently required
13	Institutions	No formal/non formal institution at village level	Forest Department, Irrigation department and Agriculture Department may play major role in these villages for IWRM practice and institutional training for better cropping and watershed management Villagers may be trained by these for better livestock and farming and non-farm activities for revenue generation Non timber forest products are very important for revenue generation these may be supported by the institutions for management Panchayat may also intervene for rain water harvesting and soil conservation work
14	Landscape	Landscape transformation	Deforestation has transformed the area. Not only by local population and contractors but development of canal and road etc. have also destructed forest Afforestation and plantation is required. Watershed management needs to be done in nearby forest land. Awareness about the changing effects of humans on landscapes, degrading forests, pressure on agriculture and irrigation need to be addressed

10.3 Institution

Forest department, Madhya Pradesh has worked for plantation in their territory. Both the villages have government and non-government intervention in terms of water conservation in Ragauli and education in Masauni. Canal water supply nearby the areas is being maintained by Irrigation Department. These villages do not have proper awareness for natural resource management and IWRM need.

10.3.1 Capacity Building

These village need training and awareness for better cropping practice, fertilizer and pesticide applications, rain water harvesting and soil conservation technology. Villagers in Ragauli are also interested in non-farm activities. However, they need awareness and technical information for the best practices. In Masauni, villagers did not showed their interest for training as the government has already given additional financial support because of minority.

10.4 Summary and Conclusions

In the area, water is really a problem. Based on the investigation different opinions were came across on the natural resources utilization and management. In the area, biasedness was observed from the government's support, which is really dangerous in sustainable development in the catchment. The principle of common government for common people should be encouraged in the region. The remedial actions or recommendation (Table 10.4) will be useful in the catchment's water resources planning and management having similar hydro-geological and climatic conditions.

References

CGWB (2010) Groundwater status of India. Central Ground Water Board, Faridabad, India

Jain SK, Agarwal PK, Singh VP (2007) Hydrology and water resources of India. Springer, New York

Rai RK, Upadhyay A, Ojha CSP, Singh VP (2011) The Yamuna river basin: water resources and environment. Water Science and Technology, vol 66. Springer, The Netherlands

Chapter 11
Lesson Learned

Abstract This chapter summarizes the lessons learned from the various case studies in the Yamuna River Basin. These learning should be adequately considered while making action plan for future Land and Water Resources Development and Management adopting the IWRM approach.

In the Yamuna River basin, because of inherent topographic and geological settings, water availability for holistic socio-economic development is really a challenge. However, in many areas, indigenous water management technology had proven a significant role to fight with this problem. It was also appeared that where the government or institutions have followed participatory approach, initiatives taken up were successful. The public participation was 100 % when government has launched any kind of IWRM programme, but public alone have not initiated any activities for their own socioeconomic development because of poor economy and uncertainty in the income from agriculture. Based on the above case studies at different parts of the Yamuna River basin, following key lesson can be learned, which makes the future Integrated Water Resources Management (IWRM) programmes successful if considered.

- **Scientific understanding:** Development of understanding on water sciences in relation to the land, topography, geology and climate to the local institution and people.
  - Hydro-meteorological and water quality monitoring.
  - Water budgeting at micro-watershed level for the planning.
  - Land use management for sustainable growth.
  - Knowledge of soil and water conservation measures or technology.
  - Sometimes construction of water harvesting structures in the catchment has largely affected the existing irrigation projects (as seen in Banganga catchment). Therefore, before construction of these structures, overall impact should be scientifically evaluated.

A. Upadhyay and R. K. Rai, *Water Management and Public Participation*,
SpringerBriefs in Earth Sciences, DOI: 10.1007/978-94-007-5709-7_11,

- **Agriculture:** Agriculture sector should be given more emphasis in the planning as in the basin approximately 90 % of population is depend on the agriculture.
  - Adequate crop planning and water allocation.
  - Irrigation efficiency: High scope of improvement in the irrigation efficiency through efficient irrigation method. However, in the basin, power supply is very irregular and has high shortage. To augment power supply, solar power system is the only solution, which should be encouraged by the government if really, has the vision for sustainable water resources management and therefore the Gross Domestic Product (GDP) growth.
  - Mixed farming need to be encouraged. There should have a balance between crop and livestock production. The livestock by-product will also help in nutrient supplement to the agricultural land so that dependency over the chemical fertilizers can be reduced.
  - No doubt MNREGA has supported the villages in reconstruction/rehabilitation and construction of water harvesting structures, roads, afforestation, etc. It has also created employment guarantee and therefore the income to the villagers. However, the continuous schemes under this act have created agricultural labour problems throughout the basin, which is really a concern in holistic and sustainable development.
- **Institutional development:** In the basin, generally institutions are not functional. Its development, revival and financial support are very important. Institution should have given authority to look after the watershed development and management. Principle of IWRM i.e. environment, economy and equity (3Es) should be adequately maintained. Based on the cast and religion, the 3Es principle should not be violated.
  - In the rural area having agriculture based economy exists, the government institutions such as agriculture department, water resources department, irrigation department and forest department should work together with an objective of sustainable growth.
- **Environment clearance and licensing**: For any new project in the watershed, cumulative impact assessment should be at the top priority. As clear from the study, huge mining work is going on in the basin, which has largely affected the groundwater. Therefore, government should not offer any fresh license or clearance without knowing the cumulative impact of the project.
- **Capacity building:** Professional training programmes should be organized to have good understanding of natural resources and their sustainable utilization. Vocational training to support the off-season income is also important.
- **IWRM:** The above components are adequately considered in the IWRM approach. Therefore, the future water resources development in the catchment should follow the IWRM approach. It should be given more emphasis to the traditional practices rather than looking for the new technology.

Index

A
Alluvial plains, 51
Anicut, 72

B
Banas catchment, 91
Banganga River, 67
Betwa, 117
Biodiversity, 53
Bundelkhand region, 117, 139

C
Central India, 139
Chambal basin, 103
Chambal River, 103
Climate, 53, 105

D
DO profile, 18
Dublin Principles, 27

E
Environment, 1
Environmental flow, 20

F
Fauna, 38
Focus group discussion, 28

I
Indian monsoon, 2
Indian River basins, 3
Integrated nutrient and pest management (INPM), 37
Integrated water resources management (IWRM), 1
Integrated watershed development project (IWDP), 39
Irrigation projects, 104

J
Johads, 72

K
Ken River, 139
Kolaba samiti, 61
Krishi Vigyan Kendra (KVK), 54

L
Land holdings, 42
Landuse, 53
Lessons learned, 153

M
Mahatma Gandhi National Rural Employment Guarantee Programme (MNREGA), 42, 121
Major tributaries, 15
Migration, 43

N
Non-timber forest products (NTFP), 37
North–East (NE) monsoon, 2

A. Upadhyay and R. K. Rai, *Water Management and Public Participation*,
SpringerBriefs in Earth Sciences, DOI: 10.1007/978-94-007-5709-7,

P
Participatory mapping, 29
Per capita utilizable water resources potential (PUWR), 5
Per capita water resources (PCWR), 4
Public participation, 115

Q
Questionnaire, 27, 28

R
Rainfed, 92
Rainwater harvesting, 98
Rajasthan, 91
River Boli, 34
River Somb, 34

S
Shivalik, 33
Sind, 117
Socio-economic, 1
Soil conservation, 44
Stages of ground water development, 17
SW monsoon, 2

T
Traditional structrures, 74

V
Vindhyan range, 117

W
Warbandi, 40
Wastewater generation, 19
Water harvesting structures, 121
Water harvesting system, 92
Water management practices, 28
Water resources, 2
Water resources status, 4
Water user association (WUAs), 27

Y
Yamuna River, 13

MIX
Papier aus verantwortungsvollen Quellen
Paper from responsible sources
FSC® C105338

If you have any concerns about our products,
you can contact us on
ProductSafety@springernature.com

In case Publisher is established outside the EU,
the EU authorized representative is:
Springer Nature Customer Service Center GmbH
Europaplatz 3, 69115 Heidelberg, Germany

Printed by Libri Plureos GmbH
in Hamburg, Germany